Marketing

Applied Process Control Instrumentation

Jean Potvin
Regulation, Automation, Metrology Inc.
(R.A.M. Inc.)

 Reston Publishing Company, Inc.
A Prentice-Hall Company
Reston, Virginia

Library of Congress Cataloging in Publication Data

Potvin, Jean.
    Applied process control instrumentation.

    1. Engineering instruments. 2. Process control.
I. Title.
TA165.P67   1985   660.2′81   84-24845
ISBN 0-8359-9238-1

10 9 8 7 6 5 4 3 2 1

Contents

1 **Boilers** 1

Elements of Combustion 1
  The Combustion Process, 1
  Requirements for Combustion, 1
  Combustion Efficiency, 2
  Factors Affecting Efficiency, 2
  Practical Aspects, 3
  Recommended Excess Air, 3

Combustion Control Systems for Gas and Oil Firing 4
  Type 1, 4
  Types 2 and 3, 8
  Type 4, 13

Drum Level Control 16
  Boiler Drum, 16
  Single Element Type, 17
  Two Element Type, 19
  Three Element Type, 21

Steam Temperature Control 23
  Convection Type Superheater, 23
  Radiant Type Superheater, 23
  Steam Attemperator, 24

Furnace Pressure Contol 27
  Furnace Pressure, 27
  Level of Operation, 28
  The Combustion Process, 28
  Furnace Draft Control System—Single Element, 29

2 **Food Industry** 33

Brine Density Control System 33
  Control System, 33

Lye Density Control System 34
  Control System, 34

High Temperature Short-Time Pasteurization 35
  Control System, 36

Ammonia Beer Cooler Control System 37
  Control System, 37
Carbonation Control System for Breweries 41
  Control System, 41
Clean-in-Place System 49
  Control System, 49
  Typical C.I.P. Cycle, 52
Retorting Operation 53
  Steam Cooking, 54
  Water Cooking, 65

3 **Water Treatment** 71
Influent Water Treatment 71
  Chemical Addition, 71
  Coagulation, 73
  Flocculation, 73
  Sedimentation, 73
  Filtration, 73
  Chlorination and Ozonation, 77
  Control Systems, 75
Water Demineralization Treatment 75
  Organic Material Trap, 76
  Cation Exchanger, 76
  Anion Exchanger, 76
  Degasser, 76
  Control System, 77
Effluent Water Treatment 80
  Primary Treatment, 80
  Secondary Treatment, 84
Sludge Concentration 92
  Digestion, 92
  Clarification, 92
  Flotation, 93
  Sludge Dewatering, 93
Sludge Disposal 96
  Land Fill, 96
  Multiple Hearth Incineration, 96
  Fluidized Bed Incineration, 97

4 **Petroleum** 99
Oil and Gas Production 99
  Classifying Hydrocarbons, 99
  Oil and Gas Traps, 100
  Production, 100

Natural Lift Wells 103
  Control Systems, 104
  Application Data, 106
Mechanical Lift Wells 106
  Valve Operation, 107
  Piston and Sucker Rod, 107
  Pump Actuation, 108
  Gas and Crude Oil, 108
  Control System, 108
The Central Battery 109
  Control System, 110
Lease Automatic Custody Transfer Unit 114
  Control System, 116
Separators 117
  Gas Content, 117
  Water Content, 118
  Types of Separators, 119
  Control System, 119
Nonassociated Gas Production 122
  Control System, 123
Gas Treatment 128
  Water Removal Process, 129
  Control System, 131
  H_2S and CO_2 Removal Process, 133
  Nitrogen and Helium Removal, 135
  Control System, 135
Absorption Oil Process 137
  Common Terms, 137
  Lean Oil Absorption, 138
  Control System, 138
Crude Oil Distillation 142
  Common Terms, 142
  Atmospheric Distillation, 142
  Vacuum Distillation, 149
Light Ends Recovery 151
  Common Terms, 152
  Light Ends Fractionation, 153
  Control System, 155
Basic Pipeline Operation 157
  Product, 157
  Pressure Loss and Pressure Contours, 158
  Pipeline Operation, 160
  Station Operation, 161
Pipeline Booster Station Control 162
  Control System, 163

5 Pulp and Paper **171**
Wood Chip Handling 171
    Control Systems, 174
Sulfate Chemical Digester 178
    The Chemical Digestion Process, 178
    Major Process Equipment Components, 178
    Cooking Methods, 178
    Batch Digesters, 180
    Control System, 180
Cooking Liquor Measuring Systems 182
Digester Blow Heat Recovery 186
    The Heat Recovery Process, 186
    Control System Objective and Classification, 186
    Control System, 188
Thermomechanical Pulping 190
    TMP Refining Stages, 190
    Refiner Stage Control System, 192
Pulp Conditioning Process 198
    Pulp Conditioning Equipment, 198
    Pulp Conditioning Control System, 201
The Paper Machine 207
    The Wet End, 207
    The Dry End, 207
    Head Box, 208
    Sheet Former, 210
    Three Pinch Roller Press, 210
    Drier, 212
    Calender, 215

6 Textiles **219**
Dyeing of Textile Yarn and Fabric 219
    The Dye Cycle, 219
    Dye Beck, 219
    Paddle Dye Machine, 220
    Beam Dye Machine, 220
    Package Dye Machine, 221
    Dye Jig, 222
    Padder, 222
    Skein Dye Machine, 223
    Jet Dye Machine, 223
    Drug Addition, 224
    Dye Liquor Temperature Control System, 224
    Dye Vessel Pressure Control System, 224
    Sequential, 224

Boil Control System For Atmospheric Dye Becks 225
   Control System, 226
High Speed Warp Slasher 228
   Control System, 229

7   **Miscellaneous Processes** 231
Gas Scrubbing Tower 231
   Gas and Particulate Scrubbing Section, 231
   The Heat Recovery Section, 231
   Control System, 233
Control System for Film Processing 234
   Control System, 234
Introduction to Reactor Control 237
   Chemical Reaction, 238
   Batch Reactor, 239
   Control System, 241
Centrifugal Compressor Surge Control 243
   Control System, 244
Basic Reciprocating Compressor Control 251
   Control System, 252

Index 255

In many nations of the world, the comma is used as a decimal point. In order to avoid confusion, this text follows the accepted scientific practice of eliminating the comma in numerical groupings to indicate thousands. This means that the number 123,456 correctly appears as 123 456 throughout this book.

Preface

There are good books on control instrumentation hardware and there are good ones on process. But very few books link process with control instrumentation. This book is a collection of process control applications involving the most popular processes found in power plants, oil refineries, pulp and paper mills, influent and effluent water treatment plants, textile mills, food processing plants, and other industries. This book will be useful to all personnel connected with process control instrumentation. Plant and project engineers, production and operation personnel, instrumentation maintenance staff, design and consulting engineers, and students of process control instrumentation will benefit from the clear, concise, and practical examples of process and instrumentation applications described here.

The content of this book was in great part provided through the courtesy of Taylor Instrument Company, "The Process Control Company."

Jean Potvin, President
Regulation, Automation, Metrology Inc.
(R.A.M. Inc.)

Acknowledgments

Appreciation is expressed to the Taylor Instrument Company for permission to use material from their Application Data Sheets and to all personnel whose names appear on those sheets. Thanks also to Paul Urbanek, of R.A.M. Inc., who spent many hours writing and editing a number of sections appearing here. Special credit goes to:

- Donald P. Benz
  Systems Engineer

- David J. Crozier
  Systems Engineer

- Kenneth R. Drayton
  Systems Engineer

- Thomas M. Fontaine
  Systems Specialist

- G. Frederic Hall
  Systems Specialist

- Harold E. Hendler
  Senior Systems Engineer

- Michael A. Mihalick, Jr.
  Senior Systems Engineer

- Paul S. Urbanek
  Instructor

- Joseph W. Zilinski
  Systems Engineer

- Robert T. Coupal
  Senior Systems Engineer

- Allan B. Dickson
  Senior Systems Engineer

- Frank S. Dryja
  Systems Specialist

- Donald T. Gregg
  Senior Systems Engineer

- Richard A. Hanson
  Systems Engineer

- David Marcus
  Systems Specialist

- Frederick E. Vandaveer, Jr.
  Senior Systems Engineer

1 Boilers

ELEMENTS OF COMBUSTION

The chemical combination of oxygen with other elements takes place around us continuously. When this process takes place rapidly, with the release of heat, the reaction is called *combustion*. Heat released by the combustion process is the major source of energy for steam generation.

THE COMBUSTION PROCESS

There are three basic requirements for combustion:

REQUIREMENTS FOR COMBUSTION

- There must exist elements which will combine with oxygen and, in so doing, will release heat.

- There must exist the proper quantity of oxygen to combine with these elements.

- There must exist a temperature above the kindling or ignition temperature of the elements.

A combustion equation is represented by the formula for the combustion of methane gas in the presence of air:

$$CH_4 + 2\,O_2 + 7\,N_2 = CO_2 + 2H_2O + 7\,N_2 + Q$$

Methane + 2 oxygen + 7 nitrogen = carbon dioxide + 2 water + 7 nitrogen + Q

where Q = 23 850 BTU/lb. (55 570 kJ/kg).

ENERGY CONTENT OF VARIOUS FUELS (GROSS HEATING VALUE)

Wood (dry)	8 640 BTU/lb.	20 131 kJ/kg.
Coal	14 094 BTU/lb.	32 839 kJ/kg.
No. 2 oil	19 600 BTU/lb.	45 600 kJ/kg.
No. 6 oil	18 640 BTU/lb.	43 450 kJ/kg.
Natural gas	23 850 BTU/lb.	55 570 kJ/kg.
Propane gas	21 564 BTU/lb.	50 244 kJ/kg.
Butane gas	21 240 BTU/lb.	49 489 kJ/kg.

ASTM FUEL OIL CLASSIFICATION

No. 1   Kerosene–stove oil
No. 2   Domestic furnace oil–light industrial oil
No. 4   Bunker A–medium industrial oil
No. 5   Bunker B–naval fuel oil
No. 6   Bunker C–heavy industrial oil

COMBUSTION EFFICIENCY

Perfect combustion implies the chemical association of fuel and oxygen in exact amounts as defined by an equation. Any deviation from perfect combustion reduces efficiency, but rarely is perfect combustion achieved.

In common fuels, there exist two basic combustible elements: carbon and hydrogen. Sulphur and other elements are present in minute quantities which burn and give off heat but their reaction is considered negligible. Each of these elements needs a certain amount of oxygen to achieve perfect combustion. The presence of a mixture of combustible fuel and an oxidant is still not enough to produce combustion. If the mixture is gradually heated, the rate of the chemical reaction increases to a point called the *ignition temperature*. At this point the reaction no longer depends on an external source of heat to produce combustion. At this state, heat is generated by the reaction faster than it is lost to the surroundings, and this results in a self-sustaining reaction.

With the continual increase in cost per unit of heat released, the objective of any combustion process is to achieve maximum efficiency.

FACTORS AFFECTING EFFICIENCY

Practically speaking, we are dealing with the flow of fuels and air into a combustion chamber. The major factor in combustion efficiency is an adequate mix so that each molecule of fuel is intimately mixed with the exact number of molecules of oxygen, and perfect combustion can occur. Fuel and air are usually fired by means of a burner. It is the function of the burner to provide the best mixing possible.

A second factor which influences combustion efficiency is the geometry of the combustion chamber or furnace. It takes time for complete mixing, so it is important that there be adequate furnace volume such that mixing and combustion is completed before the flame mass falls below the kindling temperature. Added to this factor is flame shape, which is a function of burner design and adjustment, and furnace walls' shape and heat-absorbing factors. Furnace walls have little cooling effect on the flame when the walls are made of a refractory material. To a much higher degree, the flame cools when the walls are lined with water cooled tubes.

A third factor, called dilution, results from the fact that oxygen used in combustion is obtained from air. Air is a mixture of about 21 percent oxygen and 79 percent inert gases (mostly nitrogen) with varying percentages of water vapor. Nitrogen and water vapor do not enter into the reaction, but their presence causes a lower reaction temperature. In addition, heat absorbed by dilution gases mixing with fuel is lost when the gases pass up the stack.

A fourth factor, the production of water, results from the combustion of hydrogen and oxygen. At the temperature of the reaction, water is in the vapor state. Heat released per unit of fuel consumed can be expressed in two ways. If the water remains a vapor, a low heat value is attained; if the water condenses, a high heat value is reached because the water gives up its latent heat before leaving the furnace. High heat value is rarely attained or desired, since the presence of condensed water mixed with other combustion products produces strong corrosive substances which will damage the boiler.

Perfect mixing and, therefore, perfect combustion are never achieved. Complete combustion is nearly achieved by using more air than the theoretical requirement. Thus, all available heat release is obtained from the combustibles, but at a reduced availability of that heat to the process due to the dilution effect of the additional air. The quantity of air required for complete combustion is greater than the theoretical quantity and is called *excess air*. It is expressed as a percentage above what is required for perfect combustion.

PRACTICAL ASPECTS

In burning solid fuels in a bed configuration, the fuel particle size, bed density, thickness, and shape play an important role in mixing air with the fuel. In bed burning, velocity of the air is important. Since velocity energy is a function of flow rate, mixing is generally poorer at lower firing rates.

RECOMMENDED EXCESS AIR

Solid fuels (coal, wood)	10 to 40%
Liquid fuels	Low efficiency: 7 to 16%
	High efficiency: 1 to 6%
Natural gas	Low efficiency: 5 to 10%
	High efficiency: 0.5 to 5%

Excess air level subject to burner, grate, and combustion chamber design.

Excess air varies from a maximum at low firing rates to a minimum at high firing rates. It follows that combustion efficiency is improved at a higher firing rate.

COMBUSTION CONTROL SYSTEMS FOR GAS OR OIL FIRING

Several combustion control systems exist. Four of these are outlined below and will be discussed in this chapter.

- **Type 1.** Parallel adjusted positioning type with no metering of fuel or air. Mainly used on institutional package boilers.
- **Type 2.** Parallel adjusted with full metering of fuel and air. Mainly used on boilers supplying steam to industrial processes with constant loads.
- **Type 3.** Parallel adjusted with full metering of fuel and air and also with fuel limiting. Mainly used on boilers subject to variable loads.
- **Type 4.** Any of the above with combustion analysis with manual or automatic trim. Applied where optimum performance is desired, especially when used with automatic combustion trim.

Control of the combustion process involves the addition of the proper amounts of fuel and air to the furnace in correct proportions. All combustion control systems contain the same elements. The control system can be simple or complex, depending on capacity, efficiency, and safety requirements.

Control of the combustion process has three objectives:

- Primarily, to balance continuously the heat input with the heat demand. This aspect of control is described as the *rate of firing* control.
- Secondly, to maintain this balance using the minimum amount of fuel and air. This is described as *combustion efficiency* control.
- Thirdly, to maintain safe conditions at all times in the furnace.

These objectives must be met throughout the expected operating range of the boiler.

TYPE 1

The combustion control system described here results from the fact that the fuel and air flow are adjusted in parallel from the master pressure controller. It is a positioning type system because the flows of fuel and air are not metered, but are the result of the direct positioning of the final control elements.

System Characteristics

This system is considered to be the most basic of automated combustion control systems. It is intended for use with single fuels either in gaseous or liquid form. Because of its basic simplicity, cost is minimal, but it lacks flexibility in terms of efficiency.

This is the type of system which must be calibrated to a tightly controlled set of fuel and air characteristics. Compensation for changes must be performed manually by the operator. On this control system, it is not possible to make phasing adjustments between fuel and air flow final control elements. Where the final control elements differ widely in speed of response (i.e., a fuel flow control valve for regulation of fuel flow, in conjunction with a variable speed fan for air flow control), the control system is limited to slow-to-moderate load changes.

Steam header pressure is the index of plant steam demand. In Figure 1–1, the steam header pressure transmitter, PT 101, measures the steam pressure and feeds a signal to controller PIC 101. Here a corrective signal is produced to maintain steam pressure at the set point. The controller output signal is fed to the air flow control drive FCV 102 through a fuel/air bias adjustment station, and to the fuel flow control valve FCV 101 in parallel (i.e., to both units simultaneously). A decrease in steam pressure controller output will

Rate of Firing Control

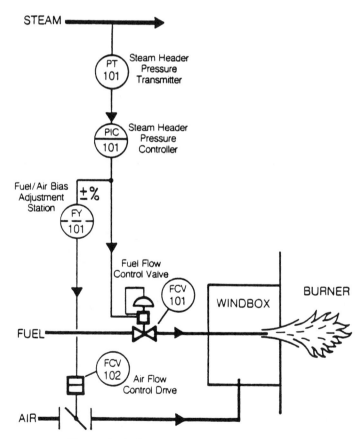

**Figure 1–1. Combustion control system—
Parallel adjusted nonmetering, Type 1.**

result in opening the air flow damper and the fuel flow control valves proportionally. The increased air and fuel flow ultimately results in greater steam production. The relationship between heat demand and the fuel flow required to meet that demand is slightly nonlinear and a function of boiler/burner efficiency in the generation and absorption of heat. This nonlinear relationship is generally not significant enough to warrant characterization of the heat demand to fuel flow relationship.

Combustion Efficiency Control

Greatest efficiency occurs with a supply of the correct amount of air to enable complete combustion of the fuel. Efficiency decreases as the air flow is increased from this correct amount. The master pressure controller output signal represents the demand for fuel and air flow quantities. This signal is fed to both final control elements (fuel valve and air drive). However, the resulting air and fuel flows are usually not in the proper proportions for efficient combustion at all load levels. This matching problem stems from two sources:

1. Inherent flow characteristics of the final control elements (see Figure 1–2). The relationship of fuel flow change is basically a linear relationship (i.e., for every pound of a given fuel burned, there is consumed a certain number of pounds of air). It can readily be seen that the first source of difficulty, which must be dealt with in matching air and fuel flow changes, is the nonlinear flow characteristics of the final control elements.

2. Characteristics of the air/fuel relationship across the expected load range. The efficiency of the combustion of fuel in a furnace

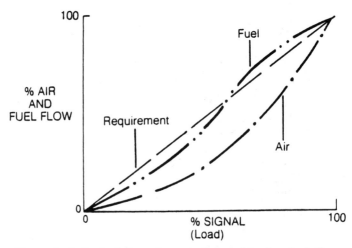

Figure 1–2. Typical flow characteristics of fuel and air flow.

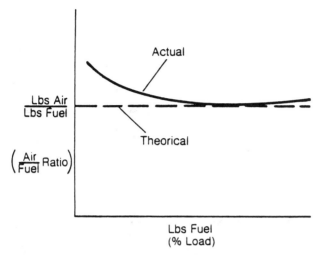

Figure 1–3. Typical variation of air/fuel ratio with load.

is heavily dependent upon the mixing of air with the fuel. At low furnace loads the velocities of the fuel and air are reduced and a mixture of a greater quantity of air per unit of fuel consumed is required (see Figure 1–3).

The final variable of the air/fuel relationship to which the system must be calibrated is shown in Figure 1–4. This is accomplished by field characterizing of the cams in the positioners of the final control elements.

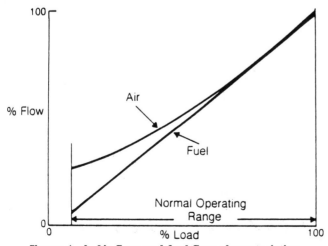

Figure 1–4. Air flow and fuel flow characteristics.

Compensation for Supply Side Load Changes

The control of fuel and air flows are not feedback systems. Any change which affects the flow through the final control element for any given controller output signal affects the air/fuel relationship for which the system was calibrated. These changes, called *supply side load changes*, include such factors as:

- Change in pressure drop relationship across the final control element from which the system was calibrated.
- Change in density or viscosity of fuel (generally accompanied by a BTU content change).
- Change in air density resulting from a change in temperature or barometric pressure.

If no provision for compensation is provided, less than optimum furnace conditions and, consequently, lower efficiency will result. A common means of compensation is to provide the operator with a bias adjustment station as shown in Figure 1–1. Compensation is accomplished by adding or subtracting an increment to the pressure controller output signal to the air flow final control element. This permits shifting the air flow, re-establishing the air/fuel relationship to the original calibration (see Figure 1–5).

INSTRUMENT LIST

Tag No.	Description
PT 101	Steam header pressure transmitter
PIC 101	Steam header pressure controller
FY 101	Fuel/air bias station
FCV 101	Fuel flow control valve (Positioner must have characterizing cam)
FCV 102	Air flow control damper (Positioner must have characterizing cam)

TYPES 2 AND 3

This section covers one control system configuration: parallel adjusted, full metering for single fuel, oil or gas, fired units.

Control System Classification

The combustion control system is classed as a parallel adjusted, full metering system. This classification results from the fact that the fuel and air flow set points are adjusted in parallel from the master pressure controller. These flows are metered and controlled by using the conventional feedback approach. Probably the most significant advantage of full metering with feedback is the close adherence to required flow even with supply side load changes. Fuel and air supply variations are compensated for before affecting each other since their response to a demand change is dependent on set

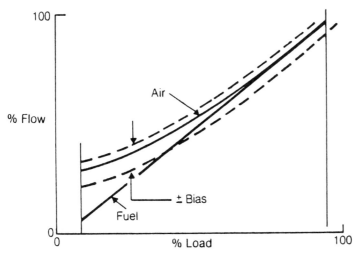

Figure 1–5. Effect of bias adjustment on air flow.

point adjustment rather than through a direct linkage system as on the type 1 parallel positioning nonmetering system.

This means that the final control elements need not be characterized to load change. While all of these advantages are realized in such features as less sensitivity to fuel and air supply changes, long-term accuracy and easier adjustments to changing conditions, they are not without disadvantages. Feedback systems of the flow type are hampered by a loss of accuracy of flow measurement at the low end of the range. However, because good, efficient firing conditions are hard to define at low loads and because the flow measurement remains repeatable in this region, this problem is minimal. Another characteristic of feedback control is lack of stability of the system. All real systems contain nonlinear elements and combustion control systems are no exception. While flow measurements are usually linearized by means of square root extractors, the final control elements can introduce stability problems on wide ranged systems. Since valve characteristics are selected, and in their manufacture closely controlled, they generally do not pose a problem. However, air modulating devices and the associated control drive and linkage can introduce nonlinear characteristics which degenerate loop stability at low flow conditions. With a little care in location of drive and adjustment of linkage, many times nonlinearity can be reduced. Otherwise, the cam in the positioner must be modified to make the system linear.

The basic control system, shown in Figure 1–6, does not contain any facilities for phasing adjustment between the fuel and air flow control system. Where these differ widely in speed of response (i.e.,

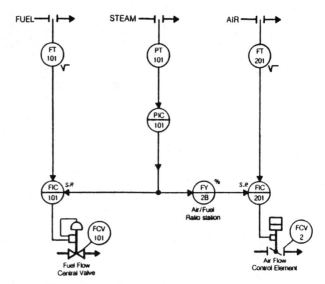

**Figure 1–6. Combustion control system—
Parallel adjusted full metering, Type 2.**

where a simple fuel flow control loop is used in conjunction with
an air flow control loop in which control of fan speed constitutes
the modulating element), the system should be limited to slow-to-
moderate rates of load change. By definition, the system is limited
to applications where load changes are slow enough for the control
system to maintain proper balance. Modification of the basic system

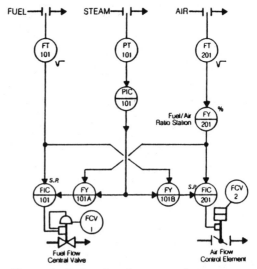

**Figure 1–7. Combustion control system—
Parallel adjusted full metering and fuel limiting, Type 3.**

to include a facility for phasing air and fuel flows is shown in Figure 1–7. The systems are cross-coupled to insure that fuel flow changes precede air flow changes on decreasing loads (i.e., for safety reasons on load changes, air supply must exceed fuel supply). This type of fuel limiting configuration permits safe and efficient use of this system on moderate-to-fast types of load change.

Rate of Firing Control (Demand). Steam header pressure is normally used to measure plant steam demand (see Figure 1–6). The steam pressure signal from transmitter PT 101 is fed to the steam pressure controller, PIC 101, called a plant master in multiple boiler systems or a boiler master in single boiler installations. PIC 101 then produces a corrective signal to maintain steam pressure at set point.

The controller's output signal is fed to the fuel flow, FIC 101, and air flow, FIC 201, controllers in parallel where it becomes the set point for fuel and air flows. A decrease in steam pressure below its set point results in increasing the steam pressure controller's output, which increases the set point of the fuel and air flow controllers. This ultimately results in greater heat release and consequently greater steam production.

Air Flow Control. Air flow is measured and linearized by transmitter FT 201. The flow signal is fed to the air flow controller, FIC 201, which produces the necessary corrective signal to maintain the air flow at its set point by adjustment of the air flow control device, FCV 2.

Fuel Flow Control. Fuel flow is measured and linearized by transmitter FT 101. The flow signal represents volume flow of fuel to the unit. Assuming a constant energy content of the fuel, this signal can be used to compute total consumed energy. The flow signal is fed to the fuel flow controller, FIC 101, which produces a corrective signal to maintain the fuel flow at its set point by adjusting the fuel flow control valve, FCV 101.

Greatest efficiency will result from supplying the correct amount of air to enable complete combustion of the fuel. Efficiency decreases as air is increased from this point. The parallel adjustment of air and fuel implies a straight line approximation of the ideal fuel to air ratio. A plot of the required fuel flow and air flow over the expected load range is shown in Figure 1–8. Note that the required air flow may be approximated quite closely from maximum firing to minimum firing point via a straight line relationship. Air flow is normally limited below this point to that required at minimum firing. This limiting is usually established by the use of mechanical stops in the control drive on the air flow final control element or by a minimum signal limiter on the control line to the drive. In

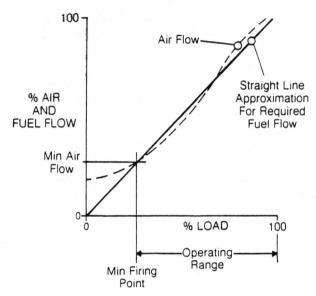

Figure 1–8. Air and fuel vs. load.

those cases where tight efficiency control is a factor, a facility to characterize air flow over the load range is recommended. In the basic system illustrated in Figure 1–6, FY 2B performs this adjustment directly on the set point of the air flow controller. In the system modified for fuel limiting in Figure 1–7, FY 201 performs the adjustment on the measured signal of the air flow controller.

INSTRUMENT LIST

	Tag No.	Description
Type 2	FT 101	Fuel flow transmitter
Parallel adjusted	FIC 101	Fuel flow controller
full metering	FCV 1	Fuel flow control valve
with fuel	PT 101	Pressure transmitter
limiting	PIC 101	Pressure controller
	FT 201	Air flow transmitter
	FY 2B	Air/fuel ratio station
	FCV 2	Air flow control damper
Type 3	FY 201	Fuel/air ratio station
Same as type 2,	FY 101A	Low signal selector (replaces FY 2B)
but with fuel	FY 101B	High signal selector
limiting system		

Oxygen trim control systems (type 4) are used to increase efficiency of fuel firing systems. Efficiency is decreased by dilution gases (mainly nitrogen) present in the air. Since only 21 percent of the air is oxygen, a slight increase in excess oxygen causes a large increase in excess air which in turn lowers the efficiency of the boiler. This decrease in efficiency is called *dilution* and is caused by nitrogen which does not support combustion but absorbs heat which escapes out the flue. To minimize wasted energy, the level of residual oxygen is measured and used to control air flow so the least possible amount of excess air exists but sufficient excess oxygen remains to complete combustion (see Figures 1–9 and 1–10).

TYPE 4

Control of the combustion process utilizing excess oxygen as a trim function can take several different forms. Industrial boiler controls are usually either mechanically linked systems or signal type control systems.

Mechanical systems use shafts and linkages to position fuel valves and air dampers. Many of these systems are not suited to an automatic control scheme because air and fuel flows are regulated by a single mechanical device which cannot be altered by pneumatic or electronic signals. In this case, only the residual oxygen content is displayed for the operator who makes manual adjustments to the fuel/air ratio.

In systems that are signal type, automatic oxygen trim control is possible by biasing the control signal fed to the air control damper. This system eliminates the use of characterizing cams since the excess air controller will produce a signal to regulate air intake until the correct value of excess oxygen is achieved.

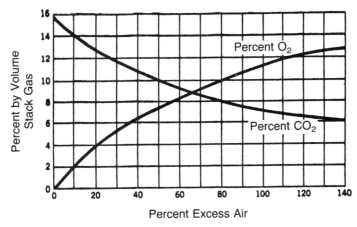

Figure 1–9. O_2 and CO_2 vs. excess air.

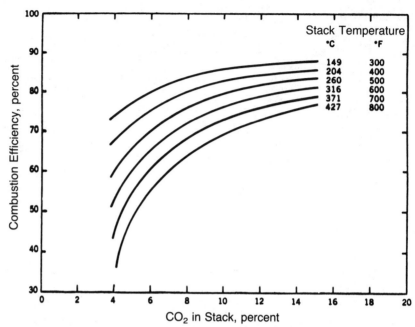

Figure 1–10. Example of effects of combustibles and diluents on efficiency.

Since excess oxygen is not a linear function throughout the entire load range, the controller set point is dependent on boiler load. There are two methods used to control automatically the set point. One uses a characterizing relay on the output of the steam header pressure transmitter, and the other uses a combustibles control system.

Oxygen Trim with Boiler Load Compensation

To compensate for the nonlinear excess oxygen level requirement, the boiler load is measured and used as a reference signal for a characterizing relay (see Figure 1–11). This relay uses the output signal from the steam header pressure transmitter to produce the air flow controller's set point. The relay is characterized to a specific boiler since each boiler differs in its excess oxygen requirement at each different load demand.

Oxygen Trim with Combustibles Analysis Compensation

Flue gases can be monitored for residual combustibles as well as residual oxygen. Combustibles analysis involves the measure of any fuel left in the flue gases (see Figure 1–12). Since residual fuel and excess air determine perfect combustion, it is logical to use these variables for control purposes.

The desired residual fuel level is set on the combustibles controller and its output is fed to the remote set point of the oxygen controller. With this control system, when load changes occur, it

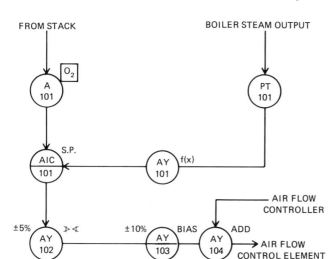

Figure 1–11. Oxygen trim control with boiler load compensation.

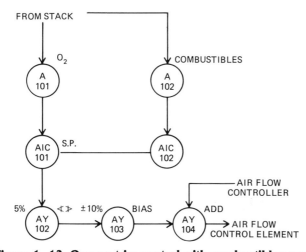

Figure 1–12. Oxygen trim control with combustibles analysis.

will maintain the same residual fuel and the excess oxygen at the optimum value like a characterizing relay but much more accurately.

Using this scheme of combustion control, the optimum operating point of each furnace can be established and combustion energy losses kept to a minimum.

The sampling system consists of a probe which takes a flue gas **Control System** sample from the stack to oxygen analyzer A 101. The oxygen analyzer feeds a signal to controller AIC 101 which produces a corrective signal which is linked to high-low limiter AY 102. This prevents the oxygen controller signal from exceeding the air flow signal. These limits are adjusted to suit different boiler requirements. After the signal passes through AY 102, it becomes the input signal for AY 103, a bias station, which will vary the control element by +/− 10 percent.

Where boiler load compensation is used, steam header pressure is measured and the signal is fed to characterizing relay AY 101 which produces an excess oxygen set point relative to boiler load.

Where combustibles analysis is used, flue gas is sampled and residual fuel measured by combustibles analyzer A 102. The measured variable is fed to combustibles controller AIC 102 which produces a corrective signal that is used as the set point for the oxygen controller. This system replaces the boiler load compensation system and offers a more efficient performance.

INSTRUMENT LIST

	Tag No.	Description
Type 4 oxygen trim system	A 101	Oxygen analyzer
	AIC 101	Excess oxygen controller
	AY 102	High-low limiter
	AY 103	Bias station
	AY 104	Summer
With boiler load compensation	AY 101	Characterizing relay
	PT 101	Pressure transmitter
With combustibles analysis compensation	A 102	Combustibles analyzer
	AIC 102	Combustibles controller

DRUM LEVEL CONTROL

BOILER DRUM   The boiler drum is located near the top of the boiler. Its main function is to provide surface area and volume for separation of steam from water. The measure of water quantity in a boiler is indicated by the level of the water steam interface in the drum.

Drum Level Control   Directly or indirectly, drum level control systems are designed to provide continuous mass balance. That is, for every pound of steam produced a pound of water is added to the drum. The liquid level in the drum must be kept within a narrow range of operation. Too high a level causes water and its dissolved solids to be carried over into the steam system. Too low a level affects the internal water treatment and re-circulation functions and may even result in tube rupture (burnout) due to lack of cooling water on the boiling surfaces.

Boiling water is a combination of steam bubbles and water and is subject to the effect of pressure, like any compressible or expandable liquid gas mixture. As such, measurement of head pressure is not always a true indication of actual water level in the boiler drum.

Swell and Shrink

There are three established drum level control approaches:

Control Classification

1. Single element type: one measured variable (level). *Application:* institutional and industrial heating plants with stable load demand.

2. Two element type: two measured variables (level and steam flow). *Application:* industrial plants with varying load, but with good feedwater regulation.

3. Three element type: three measured variables (level, steam flow and feedwater supply). *Application:* most industrial boiler systems including applications with varying feedwater supply.

Each approach is capable to a greater or lesser degree of handling variable loads and energy supply, such as varying steam demand and alternative energy content of fuel, particularly characteristic of coal, wood bark, garbage, etc.

Only in this system is the drum level measured and controlled by adding feedwater to compensate for water losses. Since the only variable affecting the position of the feedwater valve is the drum level, the system is called a single element system.

SINGLE ELEMENT TYPE

This is in contrast to a multi-element system in which steam and even the feedwater flow itself directly alter the feedwater valve position. A single element drum level control system is applied to slowly changing steam demand and to constant energy supply such as gas or oil fired boilers. In a single element system, the controller uses the drum level measurement alone as the index of load change. The measurement must therefore be a reasonably reliable indication of water quantity in the boiler. This is true only in instances where the occurrence of shrink and swell is minimal. Typical applications include institutional and industrial boilers used for space and comfort heating and for simple continuous type processes.

Drum level is measured by LT 101 and the signal transmitted to controller LIC 101 where it is compared with the drum level set point (see Figure 1–13). The controller produces the necessary corrective signal to maintain the drum level at its set point by adjusting feedwater flow. The drum level should be maintained very close to the set point value.

Control System

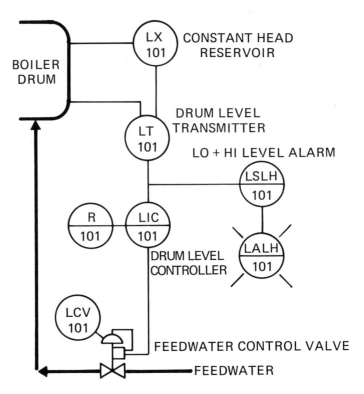

Figure 1–13. Single element drum level control system.

Drum Level Measurement. The hydrostatic head of the water between the tap near the bottom of the drum and the steam/water interface is used to measure level position of the interface. Because all drums operate under pressure, the pressure compensation leg must be tied back to the transmitter. A standard differential pressure transmitter is used for the drum level measurement. Steam in the pressure compensation leg will condense, resulting in a water-filled leg. A condensate pot is used to insure a constant head on this leg. The transmitter must be specified with an elevation adjustment feature to compensate for the wet leg head pressure.

Drum Level Control Valve. In normal service, the feedwater valve is not subjected to severe conditions. A pressure drop between 50 to 150 psig (300 to 1 000 kPag) is normal. During start-up and shutdown, up to full pump head pressure can be experienced across the feedwater valve as the drum pressure drops below header pressure. Under these conditions a high pressure drop at low flow dictates the use of at least top and bottom guided (preferably cage guided) valve plugs for lateral and vertical stability. Hardened trim is required to minimize erosion. The use of positioners is recommended on feedwater valves.

INSTRUMENT LIST

Tag No.	Description
LT 101	Drum level transmitter
LX 101	Constant head reservoir
LIC 101	Drum level controller with gain and reset response
LCV 101	Feedwater control valve
R 101	Drum level recorder, 1 pen type
LSLH 101	Level switch low and high level trip
LALH 101	Level alarm high and low level

TWO ELEMENT TYPE

This drum level system is classified as a two element type, derived from the fact that two variables, steam flow and drum level, influence the feedwater valve position. It is also known as a combination feedforward-feedback system. The load change in the form of steam flow change is fed forward as the primary index for feedwater valve position. The drum level system is a feedback system which constantly monitors the accuracy of the feedforward system and provides final control of the position of the steam/water interface in the drum.

Applications

A two element system is capable of providing close control of drum level to the set point under steady conditions as well as providing fairly accurate control of mass balance during a transient load change.

Its major restriction stems from its inability to eliminate the system of load change effects caused by feedwater variations. Performance during transient conditions permits its use on many industrial boilers. Caution should be exercised in its use on systems without reasonably constant feedwater pressure.

Flow Measurements

Steam flow can be measured by using an orifice or nozzle. Since line sizes are minimized to reduce cost of expensive high temperature and pressure piping, steam velocities run high. As such, the nozzle is widely used here because, while it exhibits less initial accuracy than the orifice (2% vs. 1%), it retains a high level of accuracy over a longer period of time in erosive service.

General Description

Load change in the form of steam flow change is measured by the flow transmitter, FT 1, whose signal is transmitted to the feedwater flow computer FY 3 (see Figure 1–14). Drum level is measured by level transmitter LT 1 and its signal transmitted to the drum level controller, LIC 1, which produces the necessary corrective signal

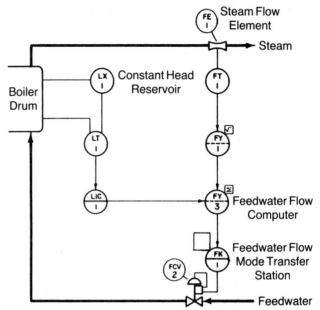

Figure 1–14. Two element drum level control system.

to maintain drum level at its set point. The controller's output is transmitted to the feedwater flow computer which combines the signals from the two variables. An output signal is transmitted to the feedwater flow control valve FCV 2. Provision for mode transfer is provided through mode transfer station, FK 101, to permit manual control of the feedwater valve.

Feedwater Flow Computer The purpose of the feedwater flow computer is to provide information on feedwater and feedback influences to the feedwater control valve. Some conditioning of the signals is also provided. The function performed by the computer is:

$$A \text{ (output)} = R(B - Kz) + C - Kb$$

where B is the steam flow signal; Kz a suppression constant; R the gain adjustment; C the drum level signal; and Kb a nulling constant. The steam flow signal is conditioned by subtracting out the suppression on the signal by constant Kz. It is then multiplied by the gain to compensate for flow range differences between the steam flow transmitter and the feedwater flow control valve. The gain is set so that for a pound of steam flow, there is a pound of feedwater flow. To permit a $+/-50$ percent bias of steam flow to feedwater flow relationship, Kb is the drum level controller's output signal at 50 percent drum level.

This drum level control system is classified as a three element type because three variables, steam flow, feedwater flow and drum level, influence the feedwater valve position.

This three element system is capable of providing close adherence of the drum level to the set point under steady conditions as well as providing tight control of mass balance during transient load changes. This control system performs well on feedwater systems which exhibit variable feed header-to-drum pressure differentials because feedwater flow is controlled. Thus, pressure drop changes across the feedwater valve are automatically compensated for before affecting the drum level. Feedwater flow control is a metered system. The linear relationship between steam and feedwater flows reduces load changes on the level portion of the system under conditions of flow change (i.e., for a pound of steam leaving the system, a pound of feedwater is added and the liquid level remains relatively constant). The performance of the three element control system during transient conditions makes it very useful for general industrial and utility boiler applications. It handles loads exhibiting wide and rapid rates of change. Plants which exhibit load characteristics of this type are those with mixed continuous and batch processing demands. Its use is also recommended where normal load characteristics are fairly steady, but upsets can occur.

Applications

The control system is the same as the two element system except for the feedwater control system (see Figure 1–15).

Control System

Flow Measurements. On feedwater flow measurement, the same reasoning as in steam flow measurement is applied although water is less erosive than steam flow. Usually if a nozzle is used on steam flow, similar factors prompt the use of the same on feedwater flow. In the feedwater flow control loop, the valve characteristic chosen is based on considerations of loop stability only. Since flow measurement is linearized, a linear valve characteristic is recommended.

Recorders and Integrators. Steam flow and feedwater flow are normally recorded on the same chart. This is done to emphasize the condition of balance (or unbalance) during transient load changes. It also provides a running check of one flowmeter against another. The difference between flows during steady state conditions represents the quantity of unmeasured flow which can indicate a possible water leak. Steam flow is normally integrated for cost ac-

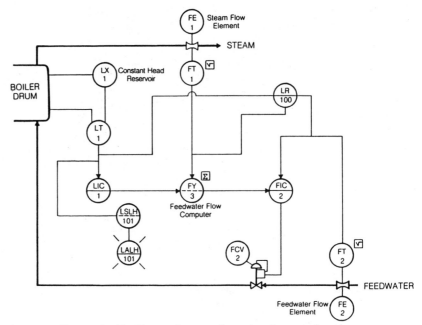

Figure 1–15. Three element drum level control system.

counting purposes. Integration of feedwater flow is recommended when a measure of total losses through blowdown and other unmeasured flow is desired. Drum level is recorded for operational purposes.

INSTRUMENT LIST

Tag No.	Description
LT 1	Drum level transmitter
LX 1	Constant head reservoir
LIC 1	Drum level controller with gain and reset response
FE 1	Steam flow element
FT 1	Steam flow transmitter
FY 3	Feedwater flow computer
FK 101	Mode transfer station
FCV 2	Feedwater flow control valve with positioner
LR 100	Drum level and steam flow recorder
LSLH 101	Level switch low and high trip (For added security, this should be independent of controller and recorder)
LALH 101	Level alarm low and high alarm
FY 1	Steam flow integrator
FT 2	Feedwater flow transmitter with square root extractor
FIC 2	Feedwater flow controller

STEAM TEMPERATURE CONTROL

Superheated steam temperature control is required in any power plant that uses turbines designed with minimum clearances and where a high efficiency operation is desired. Also, steam temperature control is usually needed by certain processes and when superheated steam is used as a catalyst. In general, any boiler that produces superheated steam must have temperature control.

One benefit gained from steam temperature control is increased boiler load range at the desired steam temperature. Turbine efficiency and life is increased by the use of superheated steam due to reduced fluid friction and erosion of the turbine. Lack of steam temperature control can cause metal failure, thermal expansion which reduces clearances in equipment, and erosion from excessive moisture particularly on the output stage of the turbine.

CONVECTION TYPE SUPERHEATER

Steam temperature control by spray attemperation is normally used with convection type superheaters designed to provide superheated steam at a given temperature at a boiler load less than its maximum rating. This type of superheater has an increasing outlet steam temperature with increasing boiler loads. Figure 1–16 shows a typical "steam temperature versus boiler load" curve for a boiler using a convection type superheater. This unit is designed to use steam attemperation for boiler loads between 70 and 100 percent steam capacity. Below 70 percent load, the final steam temperature will follow the boiler's design curve, producing steam at a temperature below its designed outlet temperature.

RADIANT TYPE SUPERHEATER

Some boilers are supplied with a radiant type superheater in series with the convection type superheater. A radiant type superheater has a decreasing outlet steam temperature with an increasing boiler load. This combination tends to flatten out the steam temperature curve over a wider load range.

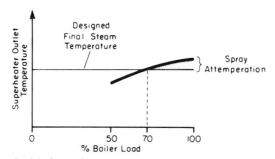

Figure 1–16. Superheater outlet temperature vs. boiler load.

STEAM ATTEMPERATOR

A steam attemperator is a device used to reduce the steam temperature. The two most common types are surface and direct contact.

Surface Attemperator

This type of attemperator isolates the steam from the cooling medium by a heat exchanger surface. Some of the common surface attemperators are the condenser, the drum, and the shell types.

Direct Contact Attemperator

This type of attemperator mixes the steam and the cooling medium, usually water, directly together. Spraying the cooling water into the steam is the most satisfactory means of changing steam temperature in the majority of applications. Atomized water provides a very large total surface area which results in a quick-acting steam temperature change when the droplets evaporate.

The spray attemperator requires high purity water with a total solids content not greater than 2.5 ppm. Boiler feedwater is used normally, but occasionally condensate is also used. The amount of spray water used for attemperation can vary between 1 and 10 percent of the boiler's total evaporation rate, depending on boiler size and superheater characteristics.

Spray Attemperator Location and Its Effect on Temperature Control

Between Superheater Stages. A spray attemperator located between two stages of a superheater will produce an average steam temperature entering the second stage which never exceeds the final desired steam temperature, allowing less expensive alloy materials to be used in the second-stage superheater. A spray attemperator at this location produces a more uniform temperature because the steam from the first stage mixes with the atomized cooling water before entering the second superheater stage. This attemperator location is usually found in superheaters that produce final steam temperatures above 850°F (455°C). The main disadvantage in this location is the complex piping between the superheater stages to the attemperator. Additionally, this location can cause large lags between the measuring point and the addition of cooling water to the attemperator, thereby calling for a multiple element temperature control system where fast rate or supply side load changes are prevailing.

Between Steam Drum and Superheater. Spray attemperators can be located between the steam drum and superheater. This is a simple piping configuration, but there is a greater chance of wet steam of reduced quality and enthalpy being delivered to the superheater. This attemperator location can give large distance–velocity lags in large superheaters.

At Superheater Outlet. When a spray attemperator is located at the superheater outlet, it may be smaller than the interstage type and a simple piping configuration can be used. The main disadvantage to this location is that the final superheated steam temperature is limited to approximately 825°F (440°C). This limitation comes from the high cost of special alloy metals required in the superheater fabrication for higher steam temperatures. Also, build-up of ash deposited by the furnace gases on sections of the superheater will cause hot spots or variations in the superheater steam outlet temperature, thereby causing a less stable final steam temperature control. When this occurs, furnace soot and slag blowers must be used more frequently.

Applications

The single element attemperator system is the simplest steam temperature control system (see Figure 1–17). It is used in boilers that have a small superheater or that have the spray attemperator located downstream of the boiler. It is also used where the rate and degree of load changes are comparatively small. In these situations, steam temperature is able to respond quickly to adjustment of the final control element.

Boilers which exhibit longer lags due to large superheater size and their attemperator location cannot respond quickly to final control element adjustments. Those with rapidly varying load demands will probably require multiple element temperature control systems.

Control System

This system measures final steam temperature by means of temperature element TE 1 and transmitter TT 1. The temperature signal is fed to the temperature controller, TIC 1, which produces the necessary corrective signal to maintain the steam temperature at its set point by adjustment of the cooling water control valve, TCV 1. Integral response in the controller is necessary because low gains are normally encountered and close adherence to the set point is required. If adjusted for the linear lags associated with the temperature element, it will be of little help in compensating for the distance velocity lag.

Steam Temperature Measurement. The type of temperature measuring element makes little difference, although a resistance or thermocouple element is generally considered faster in response to temperature changes than a capillary system. Care must be used in the selection of the thermowell to keep response time minimal. The well material should be stainless steel and have a temperature limit of approximately 1500°F (815°C). It should have good corrosion and oxidation resistance in addition to mechanical strength.

Well Location. Well location is normally downstream of the superheater and out of the flue gas area on the boiler's discharge header. In cases where the attemperator is followed by a superheater, there is sufficient mixing of the steam with the cooling medium in the

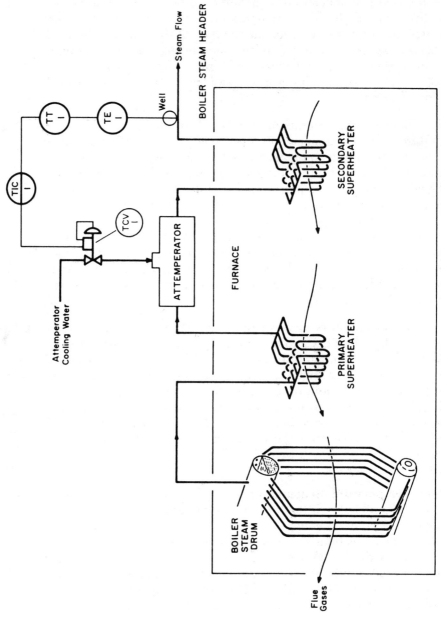

Figure 1–17. Single element steam temperature control system.

superheater so as to prevent any temperature measuring problem. When, however, the attemperator is on the outlet header, an adequate downstream run must be allowed for the steam and cooling medium to mix. Caution must be exercised not to place the well further downstream than necessary for adequate mixing, otherwise the control quality will degrade due to the occurrence of a distance–velocity lag.

Temperature Control Valve. The temperature control valve causes severe pressure drops during boiler start-up and shut-down operation. Therefore, a valve with a balanced cage guided valve plug and alloy hardened trim with a positioner is recommended.

INSTRUMENT LIST

Tag No.	Description
TE 1	Steam temperature measuring element
TT 1	Steam temperature transmitter
TIC 1	Steam temperature control
TCV 1	Steam temperature control valve

FURNACE PRESSURE CONTROL

Pressure in the furnace reflects the quantity of mass going into or out of storage in the furnace. It is an inferential measurement of mass balance. By controlling furnace pressure at a fixed value, a balance condition can be achieved between the incoming and outgoing flow.

FURNACE PRESSURE

The following elements are involved in the mass balance operation:

- Solid, liquid or gaseous fuels injected into the furnace.
- Air for supplying oxygen to the combustion process.
- The reaction gases and diluents removed from the furnace.
- Entrained solids in the flue gases and ash products from the pits.
- Air leakage into the furnace or gas leakage out of the furnace.

The volume leaving the furnace greatly exceeds that entering the furnace. This is due to the generation of gases from the reaction and the volumetric increase in inerts, nitrogen (N_2), in the air caused by the increase in temperature.

LEVEL OF OPERATION

Combustion efficiency is affected when the furnace is not operated at design pressure.

Furnaces are commonly operated under either draft or pressurized conditions.

Draft Operation

Some furnaces are not of gas tight construction. This is due to the difficulty of getting fuels such as coal and their ash products into and out of the furnace. In order to keep the combustion gases from escaping into the operating area, these furnaces are operated under draft conditions (i.e., below atmospheric pressure). To minimize the amount of uncontrolled air entering the furnace and reducing combustion efficiency, pressure is kept as close to atmospheric pressure as possible.

Pressurized Operation

Some furnaces are of all welded construction and, therefore, are gas tight. Although these furnaces may be operated at levels below, at, or above atmospheric pressure without the gas leakage or air infiltration problems associated with nonwelded construction, they are generally operated under pressure. Pressurized operation permits greater heat release due to greater throughput of fuels and air for a given furnace size.

THE COMBUSTION PROCESS

Furnace pressure operation is also classified according to the method employed for moving air and gases through the unit. The classifications are *natural draft, induced draft, forced draft,* and *balanced draft.*

Natural Draft

In a natural draft operation, the method of moving the air and gases utilizes the chimney effect. The column of hot gas in the chimney tends to rise because of the difference in density between the gases inside and outside the chimney. This causes a vacuum at the base of the column. This vacuum draws off the combustion gases and draws the combustion air into the furnace.

Natural draft operation is characterized by furnace pressure below that of the atmosphere. Control is accomplished by means of an uptake damper at the boiler outlet.

Induced Draft

This type of operation uses a fan to move the air and gases. The fan draws the gases through the boiler and in turn draws combustion air into the furnace. Since the fan is doing the work of the stack, high stacks are unnecessary. This type of operation is relatively independent of the chimney effect.

Induced draft operation is characterized by furnace pressure below atmospheric pressure. Control is maintained by the induced draft fan speed or damper opening.

This type of operation uses a fan to move the air and gases. The fan pushes the combustion air into the furnace and in turn into the stack. It is relatively independent of the chimney effect.

Forced Draft

Forced draft operation is characterized by furnace pressure above that of the atmosphere. Control is maintained by the fan speed or damper opening.

This type of operation uses both forced draft and induced draft to move the air. The mass moved by each fan must be equal.

Balanced Draft

Balanced draft operations are usually characterized by a negative pressure operation, although they can be used on positive operations. Control is maintained by regulating the air flow through the unit by one fan and the furnace pressure in the unit by the other fan.

This system is classified as a single element system because only the furnace pressure influences the position of the final control element. It is also classified as a simple feedback system. It represents the basic control system design for furnace pressure control.

FURNACE DRAFT CONTROL SYSTEM— SINGLE ELEMENT

The single element system is used on installations which are subjected to slow-to-moderate load changes. Installations characterized by large furnace volume with respect to steaming capacity, relatively simple flue path configurations, short-ducted fan installations and damper type final control elements permit use of the single element system on faster load changes.

Single element systems are found on institutional and industrial boilers used for space and comfort heating and for simple mixed batch and continuous type processes.

The furnace pressure is measured by transmitter, PT 1, and the signal is transmitted to the furnace pressure controller, PIC 1. PIC 1 then produces the necessary corrective signal to maintain the furnace pressure at its set point by adjusting the final control element (see Figure 1–18).

Control System

Furnace pressures below, above and through the zero pressure gauge are all in common usage depending upon the level of operation. Pressure transmitters with spans from 1 inch through 3 inches of water (200–800 Pa) are used for this application.

Furnace Pressure Measurement

The pressure in a furnace is characterized by a considerable amount of random pulsing. This is generally attributed to the varying rate of combustion gas generation caused by the nonhomogeneous mixing of air and fuel. Some pulsing is also the result of fan operation. This pressure variation appearing on the measurement as noise is in the range of 0.5 to 5.0 cps (Hz). Unfortunately, this noise frequency is not widely separated from the operating fre-

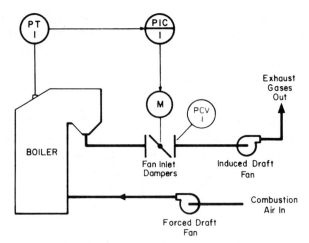

Figure 1–18. Furnace draft control system—Single element.

quency and therefore, actual load changes can be masked by the noise.

It is desirable for indication and control reasons that the noise be dampened out. If the transmitter does not contain adjustable damping, some form of external damping is desirable.

As a final control element, these various arrangements will fall under one of the following:

• Control of fan speed.

• Control of damper position.

• Any combination of these two.

The final control element can vary from a simple damper or valve operation to combinations of speed control damper arrangements.

Arrangement of forced and induced draft equipment vary widely in speed of response. It is important to know the exact final control element configuration because in some cases this configuration may be the cause of lag in the control system, which could influence the control loop's transient response.

Installation Practice

Proper selection of the point of measurement is essential to effective control system performance. The measurement tap is usually at or near the top of the furnace. It is important that a point be selected which is not influenced by changes in gas distribution or direct flame impingement. The former is important in order to attain an average measure of furnace pressure; the latter directly influences the amount of noise on the measured signal.

The tap should rise from a near vertical entrance so as to be self-cleaning. A clean-out plug should be provided for rodding the tap.

Combinations such as a fan discharge damper sequenced with fan speed modulation is a common arrangement used for efficient wide-range operation.

INSTRUMENT LIST

Tag No.	Description
PT 1	Furnace pressure transmitter
PIC 1	Furnace pressure controller
PCV 1	Final control element (damper)

2 Food Industry

BRINE DENSITY CONTROL SYSTEM

Brine solutions are used in a number of food processes. One of these is the pea sorter process. This process involves adding peas to a vat containing a brine solution with a specific density which will cause light peas to float and heavy peas to sink. This phenomenon is due to the different densities of the peas compared to that of the brine (see Figure 2–1).

The brine solution is contained in a brine storage tank in which the flow of concentrated brine is controlled by level controller LIC 100. The level signal is measured by level transmitter LT 100 on the storage tank. The level established in the storage tank must be low enough so when dilution water is added to the solution through the density controller, the brine solution does not overflow.

CONTROL SYSTEM

Brine is pumped from the storage tank into the sorter. It is also pumped into a section of vertical piping where the differential pressure transmitter PDT 100 is installed with its low input at the bottom of the pipe and its high input on the top. A change in differential of the fixed static head pressure indicates a change in density. Density controller DIC 100 receives the measured signal from PDT 100 and transmits an output signal to valve DCV 100, which controls the flow of dilution water to the brine solution storage tank.

INSTRUMENT LIST

Tag No.	Description
PDT 100	Pressure differential transmitter
DIC 100	Density controller
DCV 100	Dilution water flow control valve
LT 100	Brine solution level in storage tank
LIC 100	Storage tank level controller
LCV 100	Concentrated brine flow control valve.

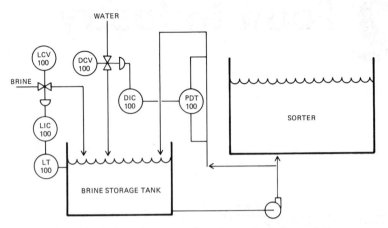

Figure 2–1. Density control system of a brine solution.

LYE DENSITY CONTROL SYSTEM

One method used to peel fruits and vegetables is immersion in a lye (sodium hydroxide) solution. This causes the peels to loosen but does not dissolve the material below the peel. In order for this process to be efficient, the lye solution's concentration must be controlled accurately since too high a density solution will dissolve the product completely and too low a density solution will not remove the peels.

Lye density can be controlled by adding concentrated lye or water. The density is also affected by changes in temperature. Since there is a constant addition of water with the product immersed in the lye solution, lye concentration can be controlled only by adding concentrated lye. The temperature will also vary the density, but it is controlled by a separate loop. Once the temperature is stable, it no longer affects the density.

CONTROL SYSTEM

The lye solution is held in a storage tank and is circulated by a pump (see Figure 2–2). The pump forces the lye solution to flow through the heater and into the peeler. The level in the peeler is constant since for all water flowing into the peeler, lye solution flows out with the peeled product. As the lye solution is drawn back into the storage tank, it flows through a sampling column where the density is measured using a bubbler differential pressure transmitter PDT 100. The signal transmitted to the density controller DIC 100 is the differential pressure between the atmosphere and the bubbler in the lye solution, representing the density of the lye solution. The density controller's output goes to control valve DCV 100 which increases or decreases as needed the flow of concentrated lye into the inlet of the pump.

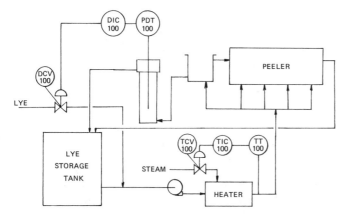

Figure 2—2. Peeler lye density control system.

Temperature of the lye solution is measured by a temperature element on the heater outlet. The measured signal is transmitted by temperature transmitter TT 100 to temperature controller TIC 100. The output of TIC 100 goes to a control valve TCV 100 which increases or decreases the flow of steam through the heater.

INSTRUMENT LIST

Tag No.	Description
TT 100	Lye solution temperature transmitter
TIC 100	Lye solution temperature controller
TCV 100	Lye solution temperature steam control valve
PDT 100	Differential pressure transmitter
DIC 100	Lye density controller
DCV 100	Concentrated lye control valve

HIGH TEMPERATURE SHORT-TIME PASTEURIZATION

An important process in the dairy industry is milk pasteurization. This involves heating milk to a temperature governed by health authorities in order to destroy any harmful bacteria present in the milk. This process is carefully monitored by government agencies. Dairies must record the temperature of the milk leaving the heater, the milk's batch number, and the position of the diversion valve which causes the milk to flow through the pasteurizer again should it be below pasteurization temperature.

Milk is supplied in vats to the pasteurizer. From the vats, the milk is pumped through a regenerator element (preheater section) and a heater element. In the regenerator element, the warm milk exiting from the pasteurizer is used to preheat the fresh raw milk coming into the pasteurizer. The milk then flows through a heater where it reaches the government regulated temperature at which point official pasteurization takes place. Since this temperature might not be attained (especially at the beginning of the batch), a temperature transmitter monitors the milk's temperature and a temperature switch actuates a recirculation valve that diverts the flow of milk back through the heater. Once the pasteurization temperature is reached, the flow of milk is directed through the regenerator and cooling section of the pasteurizer and out to the bottling line.

CONTROL SYSTEM

As the milk is pumped through the pasteurizer, the temperature measured by TT 100 is transmitted to temperature controller TIC 100 (see Figure 2–3). This controller regulates the flow of steam through the heater element, with control valve TCV 100 on the steam line, and to the temperature switch TSS 100 which controls the diversion valve on the milk flow. At the beginning the temperature is low, and so the steam flow is increased and the diversion valve is closed causing the milk to flow back into the milk vat. Once the temperature rises the recirculating valve will open and direct the flow of milk through the regenerator element, the cooling element, and then to the pasteurized milk tank.

During this time temperature recorder TR 100, using the output signal from TSS 100, plots the temperature and the position of the diversion valve. On this plotting chart is also marked the batch number. The chart is then stored for government inspection.

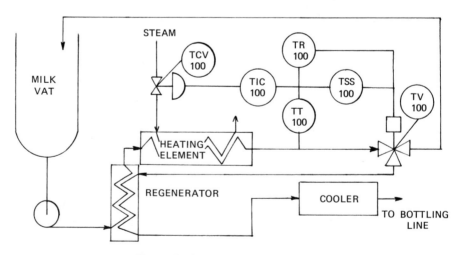

Figure 2–3. A basic HTST control system.

INSTRUMENT LIST

Tag No.	Description
TT 100	Temperature transmitter
TIC 100	Temperature controller
TR 100	Temperature and three way valve recorder
TSS 100	Temperature switch with output to three way valve
TV 100	Three way valve for recirculation of milk
TCV 100	Temperature control valve

AMMONIA BEER COOLER CONTROL SYSTEM

In a brewery the end product of the brewhouse is "Hot Wort" or unfermented beer. It is received from the brewhouse at a temperature of approximately 210°F (99°C); filtered and cooled to 50°F (10°C), and placed into the fermenter tanks. After fermentation the fermented wort (new beer) is brought to approximately 32°F (0°C) and maintained at this temperature until delivery to the packaging center. Between fermentation and delivery there are periods of storage, final filtration, adjustment of carbonation level and packaging consistency. During these processes the beer is maintained at a constant temperature by passing it through a beer cooler after each operation. If beer is allowed to attain a higher temperature during this period as compared to the final packaging temperature, it is possible that proteins will precipitate and affect beer clarity and taste.

Typically, there are three conditions under which the beer cooler will operate:

- Normal Run: Beer is flowing through the cooler and the beer temperature is being maintained at the desired value.

- Standby: A flow detecting device registers a low flow or no flow of the product through the cooler and automatically places the cooler on standby (no beer flow through the cooler). When flow is resumed, a return to normal run is automatic.

- Clean: The operator shuts the system down for the Clean-In-Place (CIP) operation.

CONTROL SYSTEM

The primary objective of the beer cooler control system is to minimize the time that the process temperature varies from the desired set point temperature and to prevent cooler freeze during the standby operation. Since the cooler's operation is essentially batch-type, there are frequent shutdowns and startups.

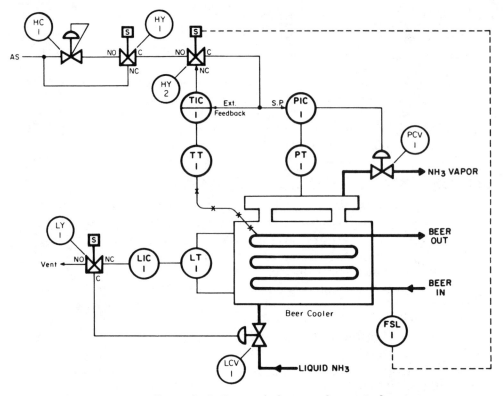

Figure 2–4. Ammonia beer cooler control system.

The beer cooler process and instrument diagram can be seen in Figure 2–4. The heat exchange medium is liquid ammonia (NH_3). As the liquid ammonia evaporates it has a cooling effect on the beer. The heat required to vaporize the liquid ammonia is equivalent to the heat energy that is extracted from the beer. The three following variables are controlled on an ammonia cooler:

• Liquid ammonia level in the main body of the heat exchanger.

• Ammonia vapor pressure.

• Beer temperature.

The beer temperature controller, TIC 1, adjusts the set point of the ammonia vapor pressure controller, PIC 1, in a conventional cascade control loop. Changes in cooling demand are seen more quickly by the ammonia pressure control loop so correction can take place almost immediately. The beer temperature transmitter, TT 1, will detect the slower temperature load changes in the beer and cause the output of TIC 1 to change. This change adjusts the ammonia pressure controller set point. As the demand for cooling continues, the liquid ammonia is constantly being vaporized. Make-up of liquid ammonia is accomplished by the level controller, LIC 1.

When a process no flow condition exists, the ammonia vapor pressure controller receives a preset manual set point from the manual loader, HC 1. Its value is higher than the normal run operation to prevent the process from freezing up. When the system is shut down, the ammonia vapor pressure controller is set to a maximum set point so that the cooler is placed at minimum cooling condition.

Normal Run Operation

Under the normal run condition of the cooler load changes occur. Most obvious is the cooling demand load changes of the beer as it enters the cooler. If the incoming beer is above the desired temperature, greater cooling is needed. This is immediately noted by an increase in ammonia vapor pressure. The pressure controller responds by opening the ammonia back pressure valve, PCV 1, until the pressure controller is satisfied. If the pressure controller is satisfied but the beer temperature controller, TIC 1, is not, the beer temperature controller will decrease the set point of the pressure controller to supply additional cooling. The cascade control system serves to maintain the most important variable, beer temperature, at the required set point. The pressure controller manipulates the variable most easily affected by process load change. The control system also corrects for changes in the supply temperature of the liquid ammonia.

Standby Operation

Under a low flow or no flow condition as detected by the inline flow switch, FSL 1, there is a noticeable change in cooling demand. When low or no flow is detected, the preset value of HC 1 is automatically put into the remote set point input of the pressure controller. This preset value is also connected to the external feedback connection of the beer temperature controller preventing its windup. The transfer from cascade to preset manual remote set point is handled by the solenoid valve, HY 2. When beer flow resumes, HY 2 returns to the normal run mode.

Shutdown Operation

The third and final mode of cooler operation is the off or shutdown operation. The system power switch is placed in the *off* position. This mode of operation is used when the cooler is not needed for an extended period of time or is shut down for a clean-in-place operation of the cooler. When the power switch is in the *off* position, HY 1 is energized, and full instrument supply pressure is placed on HY 2 (see Figure 2–5). HY 2 in turn, being de-energized, allows full instrument air pressure to the remote set point of the pressure controller. LY 1 is also de-energized in the *off* mode of the cooler. This vents the air to the air-to-open liquid ammonia valve. This vent on the signal of the liquid ammonia valve prevents the buildup of liquid ammonia in the cooler. When the clean-in-place liquid

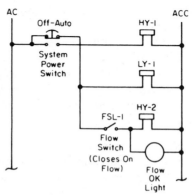

Figure 2–5. Schematic of beer cooler control system.

flows through the cooler, the trapped liquid ammonia will boil at the back pressure set by the pressure controller. From that point on, the clean-in-place liquid will virtually be maintained at its input-to-the-cooler temperature. At the end of the clean-in-place operation, the system power switch can be turned to *auto position*. This energizes LY 1 which will allow the buildup of liquid ammonia. Assuming that beer flow has not yet begun, the ammonia vapor pressure controller will have the manual set point established for the low or no flow condition and maintain that vapor pressure value. As soon as beer flow begins, the normal cascade temperature control will be in effect as previously described. It is recommended that the system (after CIP procedures) be left *off* until the beer flow is ready to begin. Operation is aimed at preventing the cooler from freezing.

Application Data

The pressure controller, PIC 1, has adjustable remote set-point stops. These are set so that the high pressure will be below the maximum design pressure for the cooler as stated by the manufacturer.

The low value is set to a minimum ammonia vapor pressure in order to protect against an excessive cooling demand by either the HC 1 manual set point or the beer temperature controller. The pressure controller is a proportional-only device with remote set point. The beer temperature controller, TIC 1, is a proportional-plus reset type controller with external feedback for anti-reset windup.

It may be required, depending on the specific application, to add alarm points to the system. These can be high and low beer temperature, and low beer flow. An adjustable two-minute time delay relay should be used with the high and low beer temperature alarm.

INSTRUMENT LIST

	Tag No.	Description
Beer Temperature	TT1	Beer temperature transmitter
Control Loop	TIC 1	Beer temperature controller
Ammonia Pressure	PT 1	Ammonia backpressure transmitter
Control Loop	PIC 1	Ammonia backpressure controller
	PCV 1	Ammonia level control valve
Ammonia Level	LT 1	Liquid ammonia level transmitter
Control Loop	LIC 1	Liquid ammonia level controller
	LCV 1	Ammonia level control valve
	HC 1	Manual loader
Solenoid Valves	LY 1	Signal switching relay
	HY 1	Signal switching relay
	HY 2	Signal switching relay

CARBONATION CONTROL SYSTEM FOR BREWERIES

The carbon dioxide (CO_2) in beer is there due to fermentation, which converts sugars to alcohol and produces CO_2 as a by-product. The brewer is interested in the CO_2 content of his beer because it affects the taste and other consumer-oriented characteristics of the beer. It also has an effect on his packaging process.

During fermentation, and all of the processing steps which follow, some of the natural CO_2 is lost from the beer. The brewer collects and uses this lost effervescence to bring the CO_2 content back up to a desirable level just before packaging. Because of inaccuracy and the time and tankage requirements of noninstrument carbonation systems, continuous measurement and control of the CO_2 content in beer is desirable and advantageous to the brewer.

The purpose of the control system is to maintain the CO_2 content of the beer at a fixed level, in spite of changes in beer flow rate, background CO_2 content, and beer or CO_2 pressure. In order to understand more easily the control concepts used in the carbonation control system (shown in Figure 2–7), it will be compared with the more conventional system shown in Figure 2–6. The cascade control system in Figure 2–6 is one which might be used to provide continuous measurement and control of the CO_2 content in beer.

CONTROL SYSTEM

The system in Figure 2–6 consists of a simple CO_2 flow control loop (FT 1, FC 1, FCV 1), the set point of which is adjusted by a volumes control loop (A 1, AC 1). If the measured CO_2 volumes

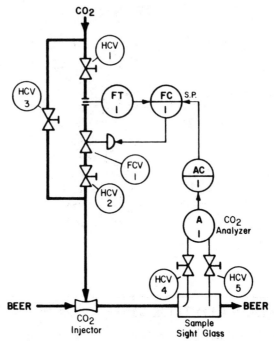

Figure 2–6. Simple CO$_2$ content control system.

are too low, the volumes control loop will increase the set point of the flow controller; if the CO$_2$ volumes are high, the set point will be lowered. Variation in CO$_2$ supply pressure or product pressure will be compensated for by the CO$_2$ flow controller. CO$_2$ volumes and flow are controlled by conventional proportional-plus reset controllers.

Experience with the carbonation control process has shown that control stability and reaction to large process upsets can be significantly improved if the discontinuous nature of the CO$_2$ volumes signal and the effective process dead time are taken into account when planning the control system. This has been done in the control system shown in Figure 2–7. The first added concept is to make the volume controller action discontinuous and synchronized with output changes from the CO$_2$ analyzer. In the conventional control system, reset action is required to eliminate offset between the set point and the process in both the flow controller and the volume controller. The reset action in the volume controller continues to make changes in the CO$_2$ flow, even during the 90 percent of the time it cannot sense the results of its changes because the analyzer output is fixed during that portion of its operating cycle. The reset action taken during the fixed output portion of the cycle may have

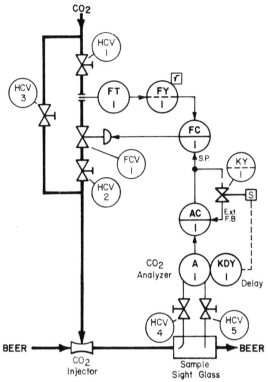

Figure 2–7. CO₂ content control system with modified feedback.

changed the CO_2 flow past the value required to maintain the desired amounts of CO_2/volume of liquid.

The synchronous-discontinuous control allows large reset set-
tings (50–200 repeats/min.) which result in crisp, not sluggish, controller action. The reset action does not continue during the 90
percent off time of the analyzer and therefore provides better control stability and better recovery after upsets. This is accomplished
by utilizing a volume controller equipped for external feedback,
and interrupting the feedback with a solenoid valve (KY 1) operated
by the cycle times of the CO_2 analyzer. When energized, this solenoid locks the pressure in the reset bellows and prevents any
change in controller output except that due to proportional action.
The controller gain is set low enough (0.1–0.25) so that proportional
action is insignificant. Reference to Figure 2–8 shows that KY 1 is
de-energized, allowing the controller to take action only during that
part of the analyzer cycle when the analyzer is presenting new
information to the control system. Controller action is thereby synchronized with the inflow of new process information, and so made
to be discontinuous.

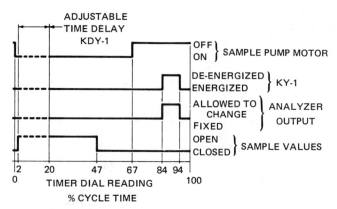

Figure 2–8. CO_2 analyzer cycle timer operation.

A second synchronizing concept has been added to the system shown in Figure 2–7, in the form of the time delay relay, KDY 1. Its purpose is to synchronize the measurement system, and thereby the control system, with reference to the particular effective dead time of the process. The system is optimized by insuring that no measurement or readjustment of CO_2 flow is made by the control system until the effect of the last adjustment has been reached and assimilated by the analyzer.

Adjustment of the time delay relay, KDY 1, lengthens the analyzer cycle so that product, with the CO_2 content dictated by the last action of the volumes controller, has time to move from the CO_2 injector to the sample inlet of the analyzer. Only after the control system has feedback regarding the results of its last corrective action can it determine the direction (increase or decrease) and magnitude of further corrective action.

Measurements **CO_2 Volumes.** There are three significant dead time lags in the CO_2 volumes measurement system. They are: firstly, the distance-velocity lag between the CO_2 injector and the sample sight glass (L_1); secondly, the distance-velocity lag of the sample lines between the sample sight glass and the CO_2 analyzer (L_2); and thirdly, the discontinuous nature of the CO_2 analyzer.

The first of these dead times (about 5 seconds long) is necessary to allow time for the CO_2 gas to go into solution before the sample is extracted. The length of piping required to provide this time can be approximated by the following equation.

$$L = \frac{\text{Max. beer flow rate}}{\text{Nominal beer line diameter}}$$

The analyzer sample taps should be located L distance downstream of the CO_2 injector.

The above equation shows that as the beer flow decreases so does the length of pipe required to produce a 5-second dead time. Since the length of pipe is fixed at L, the dead time is a function of the flow rate, and increases as the flow rate decreases. Any attempt to evaluate the effective dead time of the process must include an evaluation of this transportation lag at the current flow rate. The second dead time should be kept as small as possible by minimizing the length of the sample lines. Exceeding the recommended maximum length of 10 feet (3 meters) will degrade controllability by adding dead time to the measurement.

The third dead time is the CO_2 analyzer itself. The analyzer is a sampling instrument with a 20-second cycle. There is only a 2-second interval every 20 seconds during which the analyzer output is allowed to change.

The effective dead time of the process is a combination of these three dead times. It also includes a fourth variable—the time, relative to the start of the analyzer cycle, that a disturbance occurs in the process. Since there are two uncontrolled variables contributing to the value of the effective dead time of the process, the dead time is also a variable.

CO_2 Flow Rate. The CO_2 flow rate is measured with an orifice plate and a differential pressure transmitter. The signal is linearized using a square root extractor.

The CO_2 flow rate is controlled by a linear trim needle valve. The pressure drop across the valve and the resultant expansion of the CO_2 produce a refrigeration effect at the valve. This may cause the formation of ice on the valve body and stem. If it does, the valve and adjacent pipe should be wound with electric heating tape to melt the ice and prevent it from interfering with the smooth operation of the valve.

Control

The CO_2 flow control system quickly detects and responds to flow changes caused by CO_2 supply pressure or beer line pressure. It will also respond quickly to set point changes dictated by the primary (CO_2 volumes) control loop.

The CO_2 volumes control system includes the effective dead time described in the section titled *Measurements*. This lag makes the volumes control system slow to react. The effects of rapid changes in beer flow rate, or precarbonation level, on CO_2 content are not detected by the control system until after the effective system dead time. This means that variation in beer flow rate and precarbonation level must be slow relative to effective system dead time, or control will be less than optimum.

When properly tuned, the system will recover from set point changes and load changes in approximately ten analyzer sample

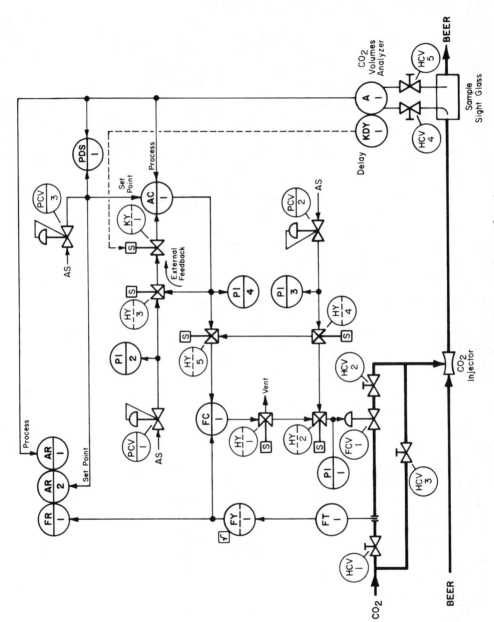

Figure 2-9. Complete CO₂ content control system.

46

times. Although this is an improvement over previous control systems, it is not fast enough to provide control of streams feeding directly into can or bottle fillers. There should be some hold up time, in the form of a tank, to average the effects of system upsets before the beer enters the fillers.

The carbonation control system, as it is installed and used, is shown in Figure 2–9. The system remains that shown in Figure 2–7 with the addition of a three-pen recorder, FR 1/AR 2; a differential pressure switch to provide deviation alarm, PDS 1; and a group of solenoid valves, pressure regulators and gauges to provide set points, manual output and reset limiting.

Application Data

In a three-pen recorder, the following variables with pen identifications as indicated are present:

- Pen AR 1—to record the volumes of CO_2/volume of beer generated by the CO_2 analyzer, A 1.

- Pen AR 2—to record the set point of the volumes controller, AC 1, which is adjusted by the pressure regulator PVC 3.

- Pen FR 1—to record the flow of CO_2 as indicated by the output of the CO_2 flow square root extractor, FY 1.

The differential pressure switch, PDS 1, measures the difference between the CO_2 volume set point and the output of the CO_2 analyzer. It provides an adjustable deviation alarm which activates a horn and flashing light when the CO_2 content of the beer varies from the set point by more than the value determined by the setting of the switch.

Interruption of external feedback to provide synchronous-discontinuous control is provided by the operation of KY 1 as previously described. An alternate source of feedback is provided through the operation of HY 3. When the carbonation control system is shut off, the reset limiting pressure determined by PCV 1 loads the reset bellows and limits reset windup. This helps, on the subsequent startup of the system, to get to the control point as quickly as possible.

The pressure regulator PCV 2 has two purposes:

- To route PCV 2 output pressure through HY 4 and HY 2 to the valve when direct manual adjustment of the CO_2 flow control valve is necessary in the case of instrument failure.

- To manipulate the set point of the CO_2 flow controller by PCV 2 output pressure, routed through HY 4 and HY 5 to the set point of FC 1, in the case of failure in the volumes control or measurement instruments.

The solenoid valve, HY 1, is operated when the system is shut off to insure that the CO_2 control valve, FCV 1, is closed by venting its valve operator to the atmosphere.

INSTRUMENT LIST

	Tag No.	Description
Simple CO_2 control system	HCV 1–HCV 5	Manual shutoff valves
	A 1	CO_2 analyzer
	AC 1	CO_2 controller
	FC 1	CO_2 flow controller using output of AC 1 for S.P.
	FT 1	CO_2 flow transmitter
	FCV 1	CO_2 flow control valve
CO_2 content control system with modified feedback	HCV 1–HCV 5	Manual shutoff valves
	A 1	CO_2 analyzer
	AC 1	CO_2 controller with external feedback signal through KY 1 switched by delay KDY 1
	KY 1	Solenoid valve
	KDY 1	Time delay relay
	FC 1	CO_2 flow controller using output signal from AC 1 as set point
	FY 1	CO_2 flow square root extractor
	FT 1	CO_2 flow transmitter
	FCV 1	CO_2 flow control valve
Complete CO_2 control system	HCV 1–HCV 5	Manual shutoff valves
	A 1	CO_2 analyzer
	AC 1	CO_2 controller with external feedback signal through KY 1 switched by delay KDY 1 and a remote set-point input

Tag No.	Description
PDS 1	Deviation alarm
PCV 1	Manual pressure regulator
PCV 2	Manual pressure regulator
PCV 3	Manual pressure regulator
KY 1	Solenoid valve
HY 1–HY 5	3-way solenoid valves
PI 1–PI 4	Pressure indicators
KDY 1	Time delay relay
AR 1	Process CO_2 recorder
AR 2	CO_2 controller set-point recorder
FR 1	CO_2 flow recorder
FC 1	CO_2 flow controller
FT 1	CO_2 flow transmitter
FY 1	CO_2 flow square root extractor
FCV 1	CO_2 flow control valve

CLEAN-IN-PLACE SYSTEM

One factor common to every processor of food and pharmaceutical products is the requirement for cleaning of all process piping and equipment after it has been used. Automatic cleaning is always more desirable than manual cleaning since it is precise and consistent in its operation and eliminates manpower.

C.I.P. (clean-in-place) cleaning techniques are by far the most efficient means of sanitizing process equipment. Piping, pumps and valves need not be taken apart each time they are cleaned. Process vessels need not be scrubbed down by hand. Some typical examples of C.I.P. applications are seen in dairies for cleaning pasteurization systems and equipment, and in both breweries and pharmaceutical plants for cleaning brew kettles, fermenting tanks and other process equipment.

CONTROL SYSTEM

Every individual food processor has his own specialized techniques for the periodic cleaning of his equipment. There is no standard C.I.P. method for all to follow. Therefore, flexibility in design and application of the control system is paramount.

C.I.P. operations are sequential by nature. Steps in the sequence generally proceed on a time basis, although there are some

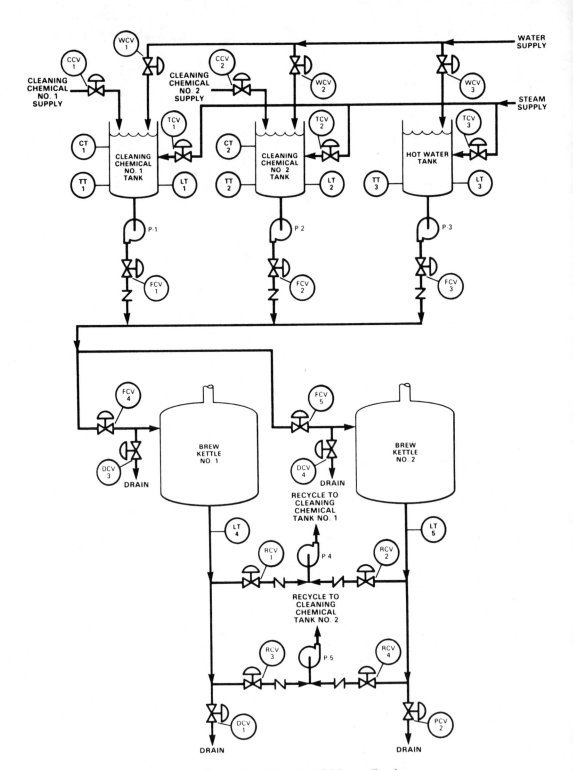

Figure 2–10. Typical C.I.P. application.

50

non-time-related events which must occur during the course of the cycle.

A computer system is ideally suited to control the C.I.P. process. The computer can monitor the status of all vessels, pumps and valves at all times to prevent costly accidents, such as caustic being accidentally pumped into a tank containing a valuable product. The computer can also time the C.I.P. operation with great precision so that the downtime of the process equipment is minimized and a more saleable product can be produced. The basic computer configuration utilizes an operator's CRT terminal, a printer for logging alarms and other important data, a disk drive for bulk memory storage, and the input/output chassis for the interface between the computer and the field devices. This configuration can be modified in many ways to customize a system for any particular process.

A typical application for a brewery C.I.P. system is shown in Figure 2–10. There are two brew kettles, tanks for two chemical cleaning solutions and hot water, and many valves, pumps and transmitters.

Each of the brew kettles has a fixed schedule for its brewing cycle and its C.I.P. cycle. The brewery can not stay on its production schedule if the C.I.P. operation overruns its allotted time. The cleaning operation must run smoothly without error or delay.

The computer monitors chemical concentrations in the cleaning solution tanks through concentration transmitters CT 1 and CT 2. It can increase the concentrations automatically by opening concentrated chemical valves CV 1 or CV 2 to supply the respective tanks. Conversely, the computer can automatically dilute the cleaning chemical solutions by adding water through valves WCV 1 or WCV 2.

The computer also monitors and controls the level and temperature in the chemical and hot water tanks. Any deviation from set point can cause an alarm to sound and be logged automatically. These controls, as well as the concentration controls, require no attention from the operators unless an alarm point is reached.

To start a C.I.P. cycle, the operator needs only to select which brew kettle is to be cleaned and push the *start* button. The computer first checks the status of all valves and pumps. It is constantly monitoring the level in each kettle by means of level transmitters LT 4 and LT 5, and so insures that the kettle to be cleaned is empty. Dumping cleaning solution into a brew would certainly destroy the product. If any of these checks show that something is incorrect, the computer sounds an alarm, prints out an alarm message for permanent record, and halts the C.I.P. cycle until the problem has been corrected.

A limit switch on the brew kettle inspection door can be connected to the computer for an emergency halt command. An unsafe condition occurs if the door is opened during a C.I.P. cycle, since the hazardous chemicals may splash out onto anyone in the area.

When all the initialization checks have been completed and proven correct, the computer begins the C.I.P. cycle. The following procedure is a typical example.

TYPICAL C.I.P. CYCLE

1. Select brew kettle for C.I.P. (e.g., Brew Kettle No. 1).

2. Push the *start* button to begin the C.I.P. operation.

3. The computer checks:
   - Status of valves on Brew Kettle No. 1, Cleaning Chemical Tanks No. 1 and No. 2, and Hot Water Tank.
   - Level in Brew Kettle No. 1 to insure that it is empty.
   - Inspection door on Brew Kettle No. 1 to insure that it is closed.

4. If the above checks are:
   - Satisfactory—process cycle starts.
   - Unsatisfactory—cycle halts, alarm sounds, printer types out data recording fault location.

5. First rinse:
   - Open drain valve DCV 1, close positive drain valve DCV 3.
   - Start hot water pump P 3, open flow control valves FCV 3 and FCV 4 to add hot water to kettle.
   - Pulse FCV 4 open and closed to get hot water blasts for thorough rinsing.
   - Rinse for specified time period.

6. First rinse drain:
   - Close hot water valve FCV 3, stop pump P 3.
   - Keep drain valve DCV 1 open for a short time to allow final draining, then close.

7. First cleaning:
   - Start Chemical No. 1 pump P 1, open valve FCV 1.
   - Start recycle pump P 4, open valve RCV 1.
   - Pulse FCV 4 open and closed to get chemical blasts for thorough cleaning.
   - Maintain Chemical No. 1 flushing for specified time period.

8. First cleaning purge:
   - Close Chemical No. 1 valve FCV 1, stop pump P 3.

- Keep Chemical No. 1 recycle pump P 4 running and valve RCV 1 open for a short time to allow final purging.
- Close valve RCV 1 and stop pump P 4.

9. Second rinse and drain:
- Repeat Steps 5 and 6 for specified time period.

10. Second cleanup and purge:
- Operate cleaning cycle and purge with Chemical No. 2, similar to Steps 7 and 8, except with valves FCV 2, FCV 4, RCV 3 and pumps P 2, P 5.

11. Third rinse and drain:
- Repeat rinse and drain cycle with hot water, the same as Steps 5 and 6.

12. Additional cleaning, purge, rinse and drain:
- Repeat Steps 7 through 11.

13. Final rinse drain:
- Open positive drain valve DCV 3.

14. Complete C.I.P. operation:
- Signal operator that C.I.P. is completed.
- Log data, times, etc. on printer.

INSTRUMENT LIST

Tag No.	Description
CCV 1(2)	Cleaning chemical supply valve
WCV 1(2,3)	Water supply valve
TCV 1(2,3)	Temperature steam control valve
TT 1(2,3)	Temperature transmitter
LT 1(2–5)	Level transmitter
FCV 1(2–5)	Flow control valve
DCV 1(2–4)	Drain control valve
RCV 1(2–4)	Recycle control valve
P 1(2–5)	Circulating pump

RETORTING OPERATION

Much of the food consumed each year is preserved by packaging it in hermetically sealed containers. This method of processing our food supply is called *canning*. One of the most important steps in the canning of foods is thermal processing, more commonly referred to as the *retorting operation*. In this operation, strict specifications as to both time and temperature must be adhered to and repeated, batch after batch, to obtain a sterile product and uniform quality.

Processing is done in batches using a cylindrical, pressurized vessel called a *retort*. Retorts are classified as either vertical or horizontal, depending on the position of the long axis. The vertical retort is loaded from the top, whereas the horizontal retort is loaded from one end. In either retort type, the canned product is cooked by one of two processes:

- Steam cooking for a product in metal containers.
- Water cooking for a product in glass jars.

The steam cooking is done in a steam atmosphere devoid of air, and the water cooking is done in steam-heated water.

TABLE 2–1. TYPICAL WATER COOKING SEQUENCE

	Vertical	Horizontal
Preheat	X	—
Load retort and close cover	X	—
Load retort and close door	—	X
Fill retort with water	—	X
Raise to cooking temperature	X	X
Cook product for a time period	X	X
Pressurized cool	X	X
Pressure relief	X	X
Drain retort	—	X
Open retort	X	X
Water cool at atmospheric pressure (if required)	X	X
Unload retort	X	X

STEAM COOKING

The steam cooking sequence is similar in either vertical or horizontal retorts:

- Load retort with the canned product.
- Lock cover or door.
- Vent the retort.
- Raise to cooking temperature.
- Cook the product for a specified time period.
- Pressure relief:
  a. For large cans, pressure cool.
  b. For small cans, relieve pressure.
- Water cool at atmospheric pressure.
- Unload the retort.

Loading, locking the cover or door, and unloading are strictly mechanical functions; therefore, additional comment is unnecessary. However, the remaining periods in the steam cook sequence require amplification because of their importance in achieving the end result.

Venting Period. Venting the retort is the first step to occur after the door, or cover, is screwed down tightly on the loaded retort. Its purpose is to eliminate the air in the retort. Steam is used to purge the air from the retort in order to obtain a true steam atmosphere. It is important to get the air out of the retort before the cooking period starts because the air, due to poor heat conductivity, acts as an insulator and prevents the canned product from being adequately cooked. Proper venting is achieved by purging for a preset time period and to a desired temperature—there are no shortcuts. Government agencies supply the processor with recommendations for proper venting.

Raise-To-Cook Temperature Period. This is the interim period after the completed vent period and raises the vent temperature to the cook temperature.

Cook Period. This period is essential in order to cook properly the product and destroy the spoilage organisms in the canned product. These organisms, called spores, are able to withstand several hours of atmospheric boiling water temperature but are destroyed in a short time if the temperature is raised to 240 to 250°F (115 to 120°C). The correct temperature and the duration of the cook period are equally important in achieving the desired objectives. Overly long cooking increases cost and may result in loss of taste or eye appeal of the product, or both.

Pressure Relief Period. The retort pressure, attained as a result of cooking at a high temperature in a closed vessel, must be relieved before the product can be safely removed. If the retort has been filled with small cans, the pressure can be reduced immediately. This reduction of retort pressure to atmospheric pressure is referred to as *blowdown*. However, if large cans are in the retort, it is necessary to lower the internal can pressure before blowdown to prevent the can from blowing up or distorting.

The reduction of the internal can pressure is obtained by circulating cooling water through the retort while the retort pressure is maintained at the value attained during cooking by introducing pressurized air. When the temperature of the cooling water leaving the retort drops to a desired value, the blowdown can take place without the threat of can damage because the product in the cans

Atmospheric Cooling

has been cooled sufficiently to lower the internal pressure to a safe level.

Water Cooling Period. In order to increase retort production, small cans are removed from the retort after blowdown, and immersed in a cooling canal prior to labelling. The larger cans, after cooling in the retort, may be additionally cooled in the same manner. The determining factor is the temperature of the product at the time it is packed in cases. This is a critical temperature because cans that are too cool may rust externally, while those that are too warm, prior to being cased and closely stacked in storage, will cool too slowly and will be damaged by "stack burn" or spoiled by creating conditions that encourage the growth of bacteria called thermophiles.

The cooling canal is a rectangular-shaped cold water-filled channel near the retort through which the can-filled baskets are pulled after they are unloaded from the retort. The length of the canal, the temperature of the water, and the speed of the conveyor determine the final can temperature. When a cooling canal is used instead of the retort for final cooling, the reduced retort time cycle increases production.

Control System The steam cook retort control systems are designed to control a cycle whose requirements are:

- Automatically controlled vent period.
- Automatically timed steam cook period.
- Pressure relief at the end of the cook period. Pressure cooling is not necessary because only small cans are processed and are strong enough to sustain atmospheric cooling.

Differentiation between vertical and horizontal retorts is necessary because of the larger volume in the horizontal retort. This greater volume requires larger control valves, etc.

The objective of the instrumentation is to provide temperature control of the product throughout the cycle. To insure that the required temperature is obtained, it is essential that the retort be properly vented and purged of air before and during the cooking period.

The retort control system can be divided into two major groups of components:

- Temperature control system used to control product temperature during cook period.
- Cycle control components used to control on/off types of functions and timing portions of the cycle.

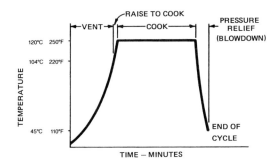

Figure 2–11. Typical time–temperature cycle for steam cook/atmosphere cool—Retort control system.

Retort Control Cycle. The control cycle is graphically shown in Figure 2–11. After the retort is loaded with can-filled retort baskets, the cover or lid is screwed down and the cycle is started. The control system will proceed through the following sequence.

Venting Period. Steam is used to purge the air from the retort. This is important to insure uniform cooking throughout the retort. Proper venting occurs when steam is blown through the retort for a preset time period and the retort temperature reaches a required value. Both time and temperature conditions must be realized before venting is considered complete.

When the cycle is initiated, the steam valve, TCV 1, and the vent valve, VCV 1, are fully opened (see Figure 2–12). Steam purges air from the retort through the vent valve until the vent timer, KS 2, finishes and the vent temperature switch, TSH 1, closes. This energizes the solenoid valve, KY 1, which applies the air pressure required to close the vent valve.

Raise-to-Cook Temperature Period. After the vent valve closes, the retort temperature very quickly reaches the preset cook temperature. When the required retort temperature is reached, the output of the temperature controller, TIC 1, to the steam valve is reduced. The steam is throttled back and the temperature controller begins maintaining the correct temperature for the cook period.

Cook Period. The reduction in controller output pressure is detected by a pressure switch, PSL 1 and its contact closure activates the cook timer, KS 1.

It is most essential that the timing of the cook period begin when the retort temperature reaches the preset value. Overcooking canned food adversely affects both taste and appearance, whereas undercooking causes spoilage.

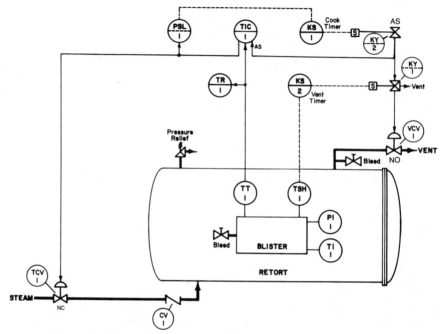

Figure 2–12. Steam cook/atmosphere cool—Retort control system.

Pressure Relief Period. At the conclusion of the cooking period, the retort pressure must be reduced to atmospheric pressure before the retort can be opened and emptied. The pressure relief period starts when the cooking timer contacts open to de-energize solenoid valve KY 2. When the solenoid is de-energized, it shuts off the air supply pressure to the controller, causing:

- Steam valve TCV 1 to close.
- Vent valve VCV 1 to open.

With the steam shut off and the vent valve open, the retort returns to atmospheric pressure. The pressure relief period is also referred to as *blowdown*.

End of Cycle. The retort is opened by the operator, who either removes the retort baskets and places them in a cooling canal, or manually introduces cooling water into the retort to cool the cans prior to removing the baskets. After the retort has been emptied, it is reloaded and the cycle is repeated.

Application Data

1. Temperature controller, TIC 1, needs a proportional response only. The proportional band should be set such that the pen

will reach the control point as quickly as possible and cut a sharp corner as it levels out at the control point. This value of proportional band normally is quite low. The steam valve, TCV 1, is oversized to provide a fast raise-to-cook temperature response.

2. The vent temperature switch, TSH 1, is factory preset to close at a temperature of approximately 220°F (104°C). The setting of the switch can be varied by a screw adjustment.

3. The vent timer, KS 2, is supplied with electrical reset and the time period is adjustable.

4. The cook period timer, KS 1, is manually reset when the cycle start button is depressed. Its time period is adjustable.

5. The hand operated bleed valves must be open during vent and cook periods to eliminate any air and noncondensibles brought in by the steam, and also to provide good circulation past the thermal element. The amount of steam condensate collected during the cycle is of no consequence and no effort should be made to remove it during the cycle.

6. The retort must be equipped with a safety relief valve that is in good working order. The safety relief valve should be adequately sized to prevent the retort pressure from going above the maximum operating pressure limit when the steam valve is wide open. Under no circumstances should a retort that does not have a safety relief valve installed be started up.

INSTRUMENT LIST

Tag No.	Description
TT 1	Retort temperature transmitter
TIC 1	Retort temperature controller
TR 1	Retort temperature recorder
TCV 1	Retort temperature control valve NC
CV 1	Steam back flow check valve
KS 1	Cook timer 0–60 min.
KS 1	Vent timer 0–30 min.
TSH 1	Vent temperature switch
KY 1	Vent timing pilot valve
KY 2	Vent timing pilot valve
VCV 1	Venting control valve NO
PSL 1	Pressure switch, low
PI 1	Retort pressure indicator 0–30 psi (0–200 kPa)
TI 1	Retort temperature indicator 170–270°F (50–150°C)

Pressure Cooling Control System

The steam cook pressure cool retort control system is designed for a cycle that consists of:

- Automatically timed vent period.
- Automatically timed and controlled steam cook period.
- Automatically controlled pressure cool period.
- Pressure relief period.

A differentiation between vertical and horizontal retorts is necessary due to the larger volume in the horizontal retort. This greater volume requires larger control valves, etc.

The objective of the control system is to control the time, temperature and pressure within the retort for the canning cycle. This is essential in order that food be properly preserved.

The retort control system can be divided into three major groups of components:

- Temperature control system used to control product temperature during the cook period.
- Cycle control components used to control on/off types of functions and timing portions of the cycle.
- Pressure control to maintain cook pressure during the pressure cool period.

The various control functions are described best by discussing them as they appear in the normal sequence of events of a standard control cycle. The control cycle is graphically displayed in Figure 2–13.

After the retort is loaded with can-filled retort baskets, the cover or lid is screwed down and the cycle is started.

Venting Period. Steam is used to purge the air from the retort to insure uniform cooking throughout. Proper venting occurs when steam is blown through the retort for a preset time period and the

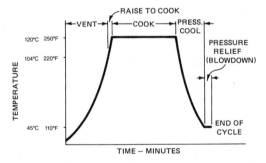

Figure 2–13. Typical time–temperature cycle for steam cook/pressure cool—Retort control system.

retort temperature reaches a required value. Both time and temperature conditions must be realized before venting is considered complete.

When the cycle is initiated by resetting the cook timer, KS 1, the steam valve, TCV 1, and the vent valve, VCV 1, are fully opened (see Figures 2–14a and 2–14b). Steam purges air from the retort through the vent valve until the vent timer, KS 2, times out, and the temperature switch, TSH 1, is closed. This energizes the solenoid valve, KY 1, which applies the air pressure required to close the vent valve, VCV 1.

Raise-to-Cook Temperature Period. When the vent valve closes, the retort temperature reaches the preset cook temperature very quickly. When the desired retort temperature is reached, the temperature controller's (TIC 1) output to the steam valve, TCV 1, is reduced. The steam is throttled back and the temperature controller begins maintaining correct temperature for the cook period.

Cook Period. Reduction in the controller's output pressure is detected by a pressure switch, PSL 1. Its contact closure activates the cook timer, KS 1. This timer controls the amount of time in which the product is held at cooking temperature.

It is most essential that the cook period timing begin when the retort temperature reaches the preset control point. Overcooking canned food adversely affects both taste and appearance, whereas undercooking can result in spoilage.

Pressure Cool Period. Some cans, because of their size and internal pressure, will distort when exposed to atmospheric pressure immediately after cooking. These cans must be pressure cooled before the retort pressure is reduced to that of the atmosphere. Cooling is done by circulating cool water through the retort while maintaining the cooking pressure in the retort. The product is cooled as rapidly as possible to prevent overcooking and to minimize retort production time.

When the cook timer, KS 1, times out, its contacts close to activate solenoid valve, KY 3, which supplies air to the pressure controller, PIC 1.

Because of different valve arrangements, the control of the vertical retort during the pressure cool period is different from the control of the horizontal retort.

Vertical Retort. In the case of the vertical retort, the pressure controller controls the retort pressure by operating the pressurizing air valve, PCV 1, and the vent valve, VCV 1. At the same time the pressure controller is activated, the cooling water valve, TCV 2, is opened. Cooling water enters the retort through this valve and exits through the vent valve.

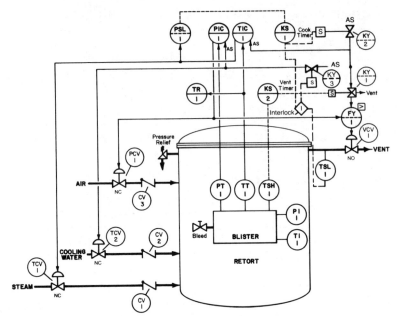

**Figure 2–14a. Steam cook/pressure cool—
Retort control system, vertical type.**

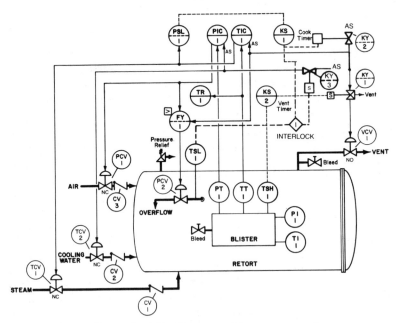

**Figure 2–14b. Steam cook/pressure cool—
Retort control system, horizontal type.**

Horizontal Retort. In the case of the horizontal retort, the pressure controller controls the retort pressure by operating the pressurizing air valve, PCV 1, and the overflow valve, PCV 2. At the same time the pressure controller is activated, the cooling water valve, TCV 2, is opened. Cooling water enters the retort through this valve and exits through the overflow valve.

To prevent loss of the cook pressure in the retort, the rate at which cooling water is initially introduced is limited. The cooling period is terminated when the temperature of the cooling water leaving the retort drops to a predetermined value. The temperature is detected by the low temperature switch, TSL 1.

Pressure Relief Period. At the conclusion of the pressure cool period, the retort pressure must be relieved before the retort can be opened and emptied. The pressure relief period is started when the cooling water temperature switch, TSL 1, shuts off the air supply to the pressure controller. The vertical retort is vented through the vent valve; the horizontal retort is vented through the overflow valve.

End of Cycle. The retort is opened by the operator, who either removes the retort baskets and places them in a cooling canal, or manually introduces additional cooling water into the retort to cool the cans prior to removing the baskets. After the retort has been emptied, the next baskets of cans scheduled for cooking are loaded and the cycle is repeated.

Application Data

1. The temperature controller, TIC 1, requires only a proportional response. The proportional band should be set such that the pen will reach the control point as quickly as possible and cut a sharp corner as it levels out at the control point. This value of proportional band normally is quite low. The temperature controller synchronization pressure is generally low because the steam valve is oversized to provide a fast raise-to-cook temperature response.

2. The pressure controller, PIC 1, requires only a proportional response. The proportional band should be set as low as possible to be consistent with the stable pressure control.

3. The vent temperature switch, TSH 1, is factory preset. The setting of the switch can be varied by a screw adjustment.

4. The vent timer, KS 2, is supplied with electrical reset. Its time period is adjustable.

5. The cook period timer, KS 1, is supplied with manual reset. Its time period is adjustable.

6. The hand-operated bleed valves must be open during venting and cooking to eliminate any air and noncondensibles brought in by the steam and, also, to provide good circulation past the thermal element. During the retort cycle some steam condensate is removed by the bleed valve while the remainder collects in the bottom of the retort below the basket of cans.

7. The cooling water temperature switch, TSL 1, is factory preset. This setting can be varied by a screw adjustment.

8. The retort must be equipped with a safety relief valve that is in working order. The safety relief valve should be adequately sized to prevent the retort pressure from going above the maximum operating pressure limit for the vessel.

INSTRUMENT LIST

	Tag No.	Description
Temperature control system	TT 1	Retort temperature transmitter
	TIC 1	Retort temperature controller
	TR 1	Retort temperature recorder
	TVC 1	Retort temperature control valve NC
	CV 1	Steam back flow check valve
Pressure control system	PSL 1	Retort pressure switch
	PT 1	Retort pressure transmitter
	PIC 1	Retort pressure controller
	PCV 1	Retort pressure control valve NC split ranged
	with VCV 1	For vertical retort
	with PCV 2	For horizontal retort
	FY 1	High pressure selector
	PVC 2	Overflow control valve NO split ranged
	with PCV 1	For horizontal retort
	VCV 1	Venting control valve NO split ranged For vertical retort
	with PCV 1	Cooling water control valve
	TCV 2	Water check valve
	CV 2	Air check valve
	CV 3	Cook timer 0–60 min.
Time cycle components	KS 1	Vent timer 0–30 min.
	KS 2	Vent temperature switch
	TSH 1	Vent timing pilot valve
	KY 1	Vent timing pilot valve
	KY 2	Cooking pilot valve
	KY 3	Retort temperature switch
	TSL 1	Retort temperature indicator
Locally mounted indicators	TI 1	170°–270°F (75°–135°C) Retort pressure indicator
	PI 1	0–30 psi (0–210 kPa)

The purposes of retort cooking of a product in glass jars are identical to those described under steam cooking in metal cans. Heated water processing differs from steam processing. A product in glass jars is processed in a water-filled retort and heated with steam. It is essential that the top jars be covered by a 4 to 6 inch (10 to 15 cm) blanket of water throughout the cycle, and that the water be agitated to provide uniform cooking. Good agitation of the heated water is attained by introducing air into the steam supply line to create a bubbling action for the complete cycle. To insure that the jar lids remain tightly sealed, the retort pressure is maintained slightly higher than the internal jar pressure. Vertical retort water cooking differs slightly from horizontal retort water cooking as indicated in Table 2–1.

WATER COOKING

Preheat Period. On the first cycle, the vertical retort is filled with water prior to loading the jars. On subsequent cycles the water used to cool down the last batch remains in the retort. The temperature of this water is too low for the next batch of hot product-filled jars and must be heated with steam to a set temperature before the jars are immersed. This operation is called preheat. Glass cannot withstand a thermal shock of more than fifty degrees. The preheat water temperature must not be more than 15 degrees above product temperature to avoid raising the internal jar pressure and causing the jar lids to pop.

Atmospheric and Pressure Cooling

Preheating in a horizontal retort is not possible because the jars must be loaded and the door closed before water can be added. The water used to fill the retort is preheated externally.

The period in which the water is heated from initial temperature to the cooking temperature is referred to as the raise-to-cook period and should be kept to a minimum to prevent overcooking.

Raise-to-Cook Temperature Period

Cook Period. Both temperature and time are critical for the same reasons as outlined in the steam cook process.

Pressure Cool Period. The internal jar pressure must be lowered sufficiently to prevent lids from popping after cooking. To do this the retort pressure is maintained as cooling water is circulated through the retort to cool the product and reduce the pressure in the jar. Temperature reduction is done quickly to stop cooking of the product.

Pressure Relief Period. After the jars have been cooled sufficiently, the retort pressure may be reduced to atmospheric pressure without danger to the containers. This relief of pressure is commonly referred to as *blowdown*.

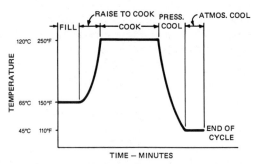

Figure 2–15. Typical time-temperature cycle for water cook.

Atmospheric Cooling Period. After blowdown and with the cover or door open, cooling can be continued in the retort by circulating cooling water through the retort. Alternately, the jars can be removed and submerged in a cooling canal, if necessary, before labelling.

Control System   The objective of the control system is to maintain the time, temperature and pressure within the retort for the entire cycle. This is most essential for the proper preservation of the food being canned.

The retort control system, shown in Figure 2–16, can be divided into three major groups of components:

- Cycle control components to manage on/off types of functions and timing portions of the cycle.

- Temperature control system to control product temperature during the cook period.

- Pressure control system to maintain the retort pressurization through the cooking and pressure cooling portions of the cycle.

The range of automatic control varies from the simplest form that provides automatic temperature and pressure control only and requires manual initiation and control of all the other steps to the more complex system that automatically programs all the steps in addition to supplying temperature and pressure control.

The system described here has automatic programming with temperature and pressure control as applied to a horizontal retort.

The sequence of events in the cycle is controlled by a programmer, KJC 1. The programmer provides both electric and pneumatic switching functions that sequentially position process control elements on a preselected schedule. Step-by-step progression through the cycle is in accord with a preselected time schedule, or in response to prescribed process conditions. A typical sequence control system for a horizontal retort provides the following operations (refer to Figures 2–15 and 2–16).

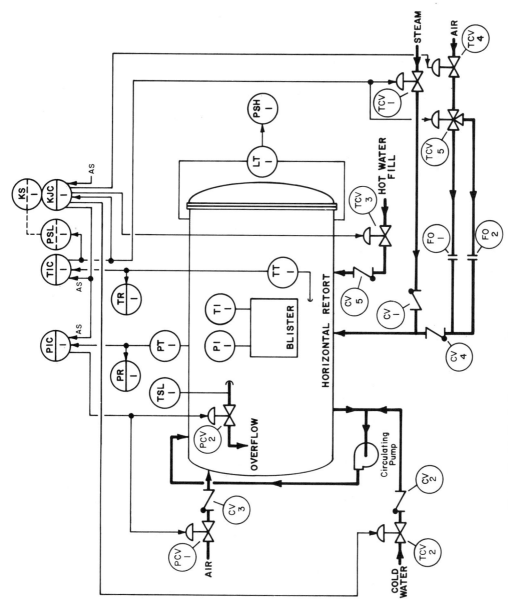

Figure 2–16. Typical water cook retort control system—Horizontal retort.

67

Preparation and Fill Retort. After the retort is filled and the end door is locked, the operator advances the programmer to the first step. The programmer opens the water fill valve, TCV 3, and checks for the correct level. At the required level a contact closure in LSH 1 advances the programmer to the next step.

Raise-to-Cook Temperature. The programmer activates the temperature controller, TIC 1, which raises the retort temperature to the cook temperature. At the same time the programmer opens the agitation air valve, TCV 4, to supply agitation air and starts the circulating pump. The pressure controller, PIC 1, is activated, establishing the required pressurization. When the retort temperature reaches its designated point, a contact closure advances the programmer to the next step.

Cook Period. The programmer times the cook period during which the temperature and pressure controllers maintain cooking conditions. At the end of the time period the programmer advances to the next step.

Pressurized Cool Period. The programmer deactivates the temperature controller to remove it from the cycle. At the same time it turns on the cooling water valve, TCV 2, and scans for the required drop in temperature of the circulating cooling water. During this period the pressure controller maintains proper pressure. When the required cooling water temperature is reached, a contact closure in the low temperature switch, TSL 1, advances the programmer to the next step.

Atmospheric Cool Period. The programmer deactivates the pressure controller. When this is done, the pressure is blown down using the overflow valve, PCV 2. The cooling water is kept circulating through the retort for a timed period. At the end of the timed period the programmer advances to the next step.

Drain Period. The programmer closes the cooling water valve, TCV 2, and the agitation air valve, TCV 4, at the same time it stops the circulating pump. The drain valve is opened to drain out the water in the retort on a timed basis. At the end of the timed period the programmer alerts the operator that the cycle has been completed.

Water Temperature Control System. The temperature of the water is measured by transmitter TT 1. The signal representing the temperature is sent to the temperature controller, TIC 1, where it is compared to the set point. The controller's output positions the steam valve, TCV 1, to maintain the required temperature.

Pressure Control System. The retort pressure is measured by the pressure transmitter, PT 1, and its signal is transmitted to the pressure controller, PIC 1, to be compared with the set point. The pressure controller positions both the overflow valve, PCV 2, and the pressurizing air valve, PCV 1, to maintain the required retort pressure.

Level Control System. The level transmitter, LT 1, measures the water level in the retort and sends its signal to a high level switch, LSH 1. A contact closure at the required water level is sent to the programmer, KJC 1, in order to close the hot water fill valve, TCV 3.

Horizontal versus Vertical Retorts. The control systems supplied for vertical retorts and horizontal retorts are slightly different. Some vertical retort systems are supplied with preheat temperature controls to preheat the water in the retort before the baskets of jars enter. Horizontal retort systems require a different approach because the retorts are loaded and the end door closed before water can be introduced into the retort. The water used to fill the retort must be heated externally.

The larger horizontal retort requires a separate pressurizing air valve in order to adequately maintain the retort pressure. It also requires a circulating pump to insure good heat transfer during cooking and good circulation or cooling water during cooling. Operation of this pump is a function of the program.

Application Data

1. The temperature controller requires only a proportional response. The proportional band should be set such that the pen will reach the control point as quickly as possible, yet cut a sharp corner as it levels out at the control point. The value of proportional band is normally quite low.

2. The pressure controller requires only a proportional response. The proportional band should be set as low as possible to be consistent with stable pressure control. If it is necessary to supply a pressurizing air valve, its operation should be staggered with the overflow valve. The pressurizing air valve should be air-to-open and the overflow valve should be air-to-close.

3. On the systems that are not supplied with a programmer, the cook timer period is adjustable. The timer is reset at the initiation of a new cycle.

4. The retort must be equipped with a safety relief valve that is in working order. The safety relief should be adequately sized to prevent exceeding the maximum operating pressure limit for the vessel.

5. The retort thermal bulb should be mounted in the bottom of the retort but not in a location that will expose it to direct steam.

INSTRUMENT LIST

	Tag No.	Description
Pressure loop	PT 1	Retort pressure transmitter
	PIC 1	Retort pressure controller
	PR 1	Retort pressure recorder
	PCV 1	Pressure control valve (split range)
	PCV 2	Pressure control valve (split range)
Temperature loop	TT 1	Retort temperature transmitter
	TIC 1	Retort temperature controller
	TR 1	Retort temperature recorder
	TCV 1	Temperature control valve
	TCV 5	Temperature control valve
System indicators	LT 1	Level transmitter
	LSH 1	Retort level switch high
	PI 1	Blister pressure indicator
	TI 1	Blister temperature indicator
	TSL 1	Temperature switch low
	PSL 1	Pressure supply low
Programmable controller	KS 1	Sequencer switch
	KJC 1	Sequence programmer
	TCV 3	Hot water control valve
	TCV 4	Air control valve
	FO 1(2)	Flow orifices
	CV 1–5	One way valves

3 Water Treatment

INFLUENT WATER TREATMENT

Water pumped from a river or any other body of water is rarely potable. Due to the presence of many pollutants, some natural and others unnatural, before this water can be used for human consumption or certain industrial processes, treatment is essential (see Figure 3–1).

Water treatment procedures vary in each plant. Large pollutants are removed with coarse screens. Chemicals are then added to increase the efficiency of coagulation and flocculation where the pollutant particles are removed by sedimentation. Next, the water is put through sand filters and then chlorinated. Further treatment such as ionic exchange, ozonation and fluorination may be part of the process.

The different steps used in water filtration affect the final quality of the output water, and a certain standard must be maintained throughout the process. This would include testing for dissolved oxygen, ammonia, organic matter, color, suspended matter, conductivity, chlorine, hardness, pH, turbidity (clearness), and dissolved carbon dioxide. These factors are either monitored constantly with continuous analyzers or are laboratory analyzed at regular intervals.

Water pumped from a river or lake flows through screens which remove large pollutants from the water. These pollutants can be anything from dead fish and weeds to car parts. The first stage of water treatment is clarification. In the clarifier, a variety of chemicals can be added, which we will look at individually. The point of the clarifier is to remove micro-organisms, color and suspended pollutants. This is accomplished by coagulation and flocculation.

CHEMICAL ADDITION

The chemicals which can be added include chlorine, alum, and caustic soda. Chlorine is added as an oxidizing agent. It will oxidize organic matter and iron which disrupts the flocculation process. To

71

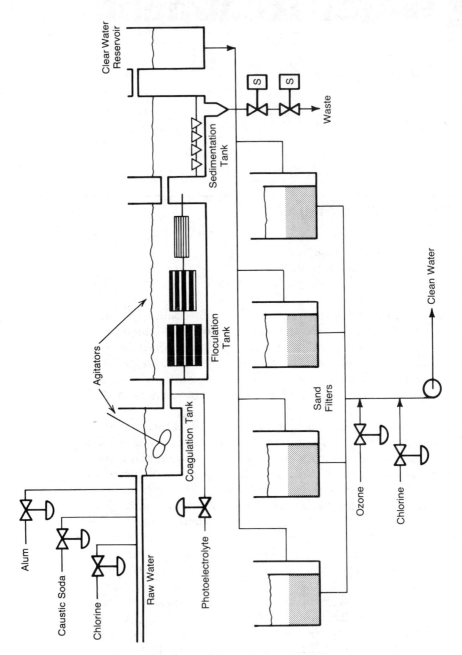

Figure 3–1. Overall view of a water treatment plant.

accomplish coagulation, a coagulant is added to the water. This coagulant can be alum, also known as aluminum sulphate. Since the mixture of alum and chlorine lowers the pH of the water, the pH value is increased with the addition of caustic soda at the input of the clarifier.

The process of coagulation involves occupying the neutral charges surrounding the pollutant particles in the water with alum particles which neutralize the charges and break apart the pollutant particles. This process necessitates the rapid stirring of the water in order to mix the alum and the water evenly. Once the water has been mixed, the next process is flocculation.

COAGULATION

In the flocculation process, a flocculant aid can be added to the water. This can be a polyelectrolyte, which causes the weighed down flocs to fall to the bottom. This process requires a slow and constant stirring of the water so all the particles come into contact with the alum and the alum is prevented from breaking away from the charges.

FLOCCULATION

The water flows from the flocculation tank into a sedimentation tank where the velocity of water flow is very slow and the flocs remain on the bottom of the tank. The bottom of the tank usually contains scrapers that remove the floc sediment and force it out the drain. From the sedimentation tank the water flows to the clear water reservoir. Here the water is stored, awaiting the sand filters.

SEDIMENTATION

Sand filters are made up of different layers of sand and rock gravel (see Figure 3–2). Fine sand is on the top layer and the size of the sand and rock gravel increases with depth. In some cases a layer of anthracite is put on top of the fine sand. Anthracite is a type of coal that is lighter than sand but larger. What occurs is the formation of a fine layer of pollutants on top of the sand. This layer then blocks the flow of water through the sand, causing the need for more frequent backwash cleaning, but anthracite is larger than the fine sand and traps the particle matter but still permits the flow of water. Prevention of this problem was also attempted by using a larger size of sand over the fine layer, but during the backwash cycle the larger and heavier sand would find its way below the fine sand layer. The depth of the sand layers varies from 24 to 36 inches (60 to 100 cm) depending on the nature of the influent water.

FILTRATION

Water is poured into the sand filter from the side and flows through by force of gravity. There is a level controller in the basin to control influent water flow. This level control loop is made up of a bubbler system and a level controller with a control valve on the influent line.

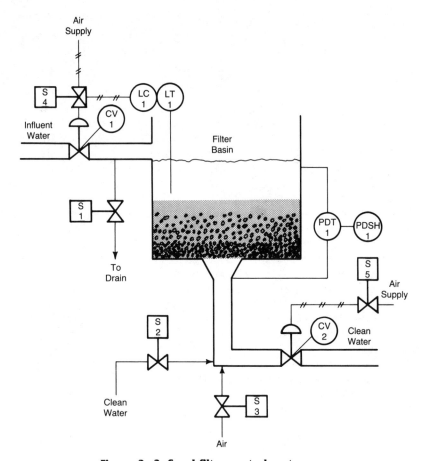

Figure 3–2. Sand filter control system.

There is also a control system to monitor the condition of the filter. This system involves a differential pressure transmitter that measures the differential pressure between the head and the outlet of the basin and a controller which switches the state of solenoid valves. When the differential pressure between the head and the outlet of the reservoir reaches a predetermined value, the sand filter is cleaned. This is performed by shutting off the inlet and the outlet water line valves and injecting a mixture of air and clean water through the base of the filter so as to loosen the layers of sand and wash away the pollutants collected there. The reverse flowing water is discharged to waste.

Once the water has been filtered, it is stored in a clearwell. After storage, it can be ozonated and/or chlorinated before being either pumped to another water treatment stage or used for consumption.

Ozonation is a process involving the addition of ozone (O_3) to the effluent water of the filters. The addition of ozone causes a reaction in the water which destroys bacteria. Ozone can be produced at the filtration plant by the ionization of air using electrodes at a high a.c. potential, causing air to dissociate and then recombine to form ozone.

CHLORINATION AND OZONATION

Prechlorination is controlled by a simple flow loop. The chlorine level is measured on the effluent line from the clarifier and controlled by adjusting the flow of chlorine into the influent water line to the clarifier.

CONTROL SYSTEMS

The amount of alum and polyelectrolyte to be added is determined by laboratory testing and adjusted to match the correct level in proportion to the water flow. In winter, summer or during stormy conditions, the quality of water can change considerably, requiring more or less chemical additives.

The filters use a combined logic switching and analog control system. The analog control system involves monitoring the differential water pressure between the head and the outlet of the filter. When the pressure reaches a preset value, the logic switch closes the valves at the inlet and the outlet of the filter, opens the drain valve next to the inlet, and opens the pressurized air line and the clean water valves. These are kept open for a predetermined amount of time dependent on the size of the filters (usually half an hour). When the clean water, air, and drain valves are closed, the inlet and outlet valves are opened to resume the filtering process.

INSTRUMENT LIST

Tag No.	Description
LT 1	Filter basin level transmitter
LC 1	Filter basin level controller
CV 1	Influent water control valve
CV 2	Clean water control valve
PDT 1	Differential pressure transmitter
PDSH 1	Differential pressure switch high trip
S 1–5	Control solenoid valves

WATER DEMINERALIZING TREATMENT

Boilers must run at the highest possible efficiency level at all times. They cannot be shut down for even the shortest period of time. Because of this, boiler maintenance is of critical importance to insure a constant supply of energy.

When water is turned to steam in a boiler, any foreign material in the water will be precipitated to the bottom of the boiler, along the walls of the boiler tubes, and in the distribution system piping. This causes an insulating layer which will decrease heat transfer efficiency as well as constrict the piping passages and restrict steam flow. One way of maintaining a clean heat transfer surface is by adding cleaning chemical products, which are expensive, or by purifying and demineralizing the boiler water supply.

The water demineralizing treatment system is made up of three stages consisting of an organic material trap, a highly acidic cation exchanger, and a highly basic anion exchanger. These exchangers remove positively and negatively charged ions by attracting oppositely charged ions. Clarified and filtered water first passes through the organic material trap to remove any material that remains after filtering. Following this, the water passes through the cation exchanger and the anion exchanger.

Following demineralization, the water is further processed by a degassing unit to remove oxygen and any other gases contained in the water. Since these gases will cause premature corrosion in the boiler and the piping around it, this stage is just as important as all the other water treatment stages.

ORGANIC MATERIAL TRAP

The trap contains anionic resin similar to the anion exchanger resin, but with a higher porosity than normal. This resin traps organic matter and gases, like CO_2, that would otherwise irreversibly contaminate the resin in the final anion exchanger. Through the trap, the water's sulphites and bicarbonates will be exchanged for chloride ions, and the trapped resin will be regenerated by the addition of brine.

CATION EXCHANGER

The cation exchanger takes all the cations in the water, such as calcium, magnesium and sodium, and replaces them with H^+ ions. Once all the H^+ ions are depleted, the resin must be regenerated. This is accomplished by the addition of sulphuric acid.

ANION EXCHANGER

In the anion exchanger, the resin exchanges OH^- ions for any anion that comes in contact with the resin. These could include sulphates, chlorides, silica and carbonic gas. The anion resin is regenerated by caustic soda. Also, all the acid formed in the cation exchanger now reacts with the base in the anion exchanger to form pure neutral water.

DEGASSER

In this unit, the demineralized water flowing towards the boiler feed pumps is degassed. This means that all gases present in the water are removed by adding steam under pressure to the tank and controlling the damper on the exhaust vent to limit the amount of steam leaving the tank with the gases. From here the water is pumped to the boilers.

The demineralization water treatment system is controlled by a logic controller (see Figure 3–3). The only PID controller in the system is a pressure controller at the water inlet of the system which maintains constant water pressure.

Throughout the system there are logic controller activated valves. These can shut off the flow depending on the status of the flow transmitters and the function being performed at the time, such as that of exchanging, regenerating the resin or treating the water.

At the intake of the organic material trap, the flow is measured and displayed on an indicator. The trap is equipped with a pressure differential switch which is used by the logic controller to stop the flow of water and clean the exchanger by backwashing.

When the operator finds that the resin is no longer capable of exchanging ions, the flow of water is cut off. The resin is then regenerated by the addition of the proper reaction material to the trap. To determine the efficiency of the resin, water conductivity is measured at the inlet and outlet of the trap. If the differential is within the preset norms, the resin is considered in good condition. On occasion, due to wear and use, new resin must be added to the trap. This is determined by a calculation based on the quantity of treated water and a visual verification.

In the cation and anion exchanger, there are several valves used to control the flow of both the water and the regenerating agent. When the differential pressure transmitter indicates a high differential, the backwash sequence is engaged. Here, the flow of water to the tank is interrupted and pressurized air is injected into the tank causing the resin pellets to separate and the attached ions to disengage. Water from the demineralized tank is used to backwash the exchanger and flush foreign material to the drain. This stage completed, a regenerating agent is added to the exchanger to renew the resin.

To purify larger amounts of water, the system can be multiplied several times by having two or more sets of exchanger tanks with a common inlet and outlet. This guarantees that the whole system does not shut down when backwashing or regenerating one set of exchanger tanks. The flow of demineralized water is then collected in a demineralized water tank so the flow of water can be constant and ready for any sudden boiler demand.

The logic controller also controls the various regenerating agent tanks and the transfer of agents from the main tank to a daily supply tank. Level control in the tanks is in on/off switching performed once a day. The switching also includes motors used to feed the regenerating agents to the resin tanks.

The degasser is equipped with a pressure controller on the tank that manages the flow of pressurized steam into the degasser (see

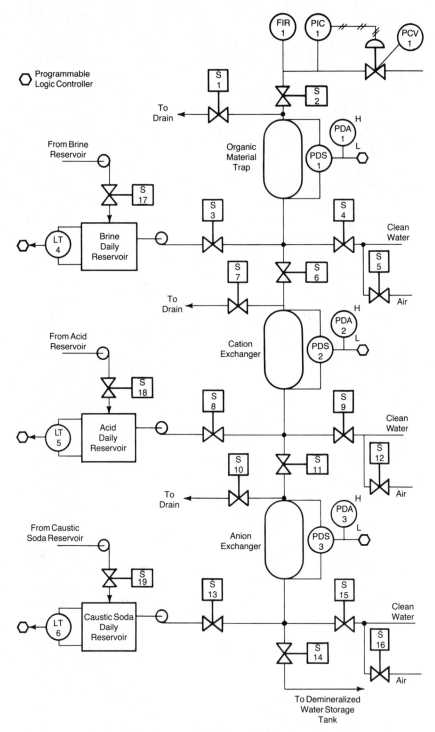

Figure 3–3. Demineralization stage control system.

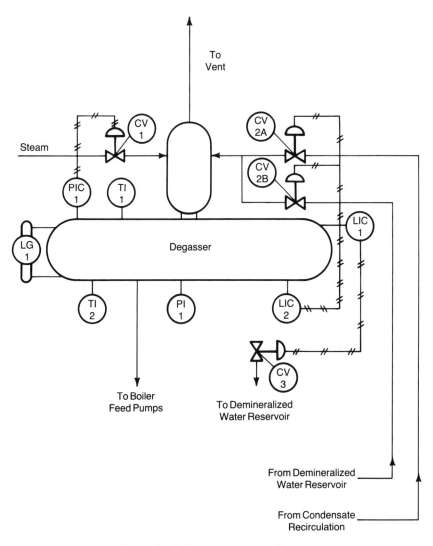

To
Vent

Steam

Degasser

LG
1

PIC
1

TI
1

CV
1

CV
2A

CV
2B

LIC
1

TI
2

PI
1

LIC
2

CV
3

To Boiler
Feed Pumps

To Demineralized
Water Reservoir

From Demineralized
Water Reservoir

From Condensate
Recirculation

Figure 3–4. Degasser control system.

Figure 3–4). It also has a low level controller which operates the
valve on the water inlet to the degasser and a high level controller
which opens a drain line that recirculates the excess water to the
demineralized water tank. There is a temperature indicator and
pressure indicator on the tank as well as a sight glass for level
indication.

INSTRUMENT LIST

	Tag No.	Description
Demineralization stage control system	PIC 1	Influent water pressure controller
	CV 1	Influent water control valve
	FIR 1	Influent water flow recorder
	PDS 1	Organic material trap differential pressure switch
	PDA 1	High and low differential pressure alarm
	PDS 2	Cation exchanger differential pressure switch
	PDA 2	High and low differential pressure alarm
	PDS 3	Anion exchanger differential pressure switch
	PDA 3	High and low differential pressure alarm
	LT 4	Brine daily reservoir level transmitter
	LT 5	Acid daily reservoir level transmitter
	LT 6	Caustic soda daily reservoir level transmitter
	S 1–19	Solenoid control valves Programmable logic controller
Degasser control system	LIC 1	Degasser level indicator controller
	CV 1	Level control valve
	LIC 2	Degasser level indicator controller
	CV 2a–2b	Level control valve
	PIC 1	Degasser pressure controller
	CV 1	Pressure control valve
	TI 1	Degasser steam temperature indicator
	TI 2	Degasser water temperature indicator
	PI 1	Degasser pressure indicator

EFFLUENT WATER TREATMENT

PRIMARY TREATMENT   Primary waste treatment is the first major treatment in a wastewater purification process. Its function is the removal of floating and settled solids which are contained in the wastewater. Figure 3–5 is a process flow diagram outlining essential unit operations. Floating

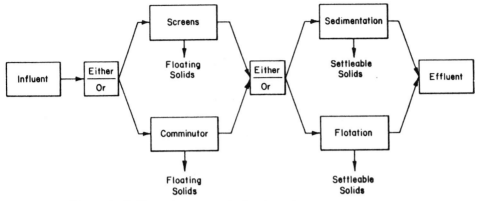

Figure 3–5. The primary waste treatment process.

solids are usually removed first because they present a potential source of damage to pumps and other equipment.

The removal of floating solids is generally accomplished by coarse screens which are kept free of collected material by rake mechanisms or by backwashing with water, air or steam. An alternate method involves the use of a comminutor which screens out some floating solids and grinds up the rest so they can be removed by subsequent treatment processes.

Settled solids are removed by either sedimentation or flotation processes. Of the two, sedimentation is the most common operation. It utilizes time and gravity to attain liquid-solid separation. Where the sedimentation basin or tank is well designed, inlet and outlet locations are placed so that short circuiting cannot occur and the transport time from inlet to outlet is sufficient to allow settled solids to precipitate into the solids removal zone. Under these conditions, sedimentation will remove 50 to 60 percent of the settled solids.

When a liquid contains relatively high concentrations of suspended particles such as those found in pulp and paper wastes, the settling rate becomes extremely slow (see Figure 3–6). Unless space is available for a large sedimentation lagoon, effective liquid-solid separation may not be possible.

Flotation is a method used to achieve liquid-solid separation where high concentrations of settled solids are involved. Separation is accomplished by the injection of minute air bubbles into the liquid which become attached to the solids, causing them to float to the surface where they are easily skimmed off. Several advantages of the flotation process are:

- It works well on liquids containing high concentrations of solids.

- It retains less liquid in the removed solids.

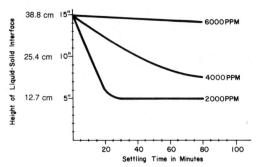

Figure 3–6. Settling rate.

- It removes a higher percentage of the settled solids than can be removed by sedimentation, and requires a smaller treatment basin.

The primary disadvantage of the flotation process is that it requires mechanical components and is more complex and higher in initial cost than the sedimentation process.

Public demand for improved waste treatment has led to the development of specialty chemicals which aid flocculation of the solid particles. These chemicals are being used in both processes to improve the removal of settled solids during primary treatment.

Control System

Primary Treatment Screens. As floating solids accumulate on a screen, liquid flow is restricted and a difference in liquid height occurs between the upstream and downstream sides of the filter. When this differential reaches a value which either the filter manufacturer or the plant operator feels to be an allowable limit, the filter must be cleaned.

Vibrating screens, drum screens and some types of traveling screens are continuously in motion and require little control other than alarm functions. The most common screen, however, is a traveling screen whose rotation and cleaning are manually or automatically controlled by measuring the differential water height across the screen and operating both rotation and cleaning methods as required by filter loading.

As shown in Figure 3–7, differential water height across the screen is sensed by a differential pressure transmitter using bubble tubes both upstream and downstream of the screen. Output from the transmitter is connected to a recorder having an adjustable high alarm contact. When the differential reaches the alarm setting, an auxiliary timer is energized and the screen drive motor and backwash solenoid valve are energized for the period of time set on the auxiliary timer. If the differential across the screen has not dropped below the alarm point at the end of the preset time period, the high

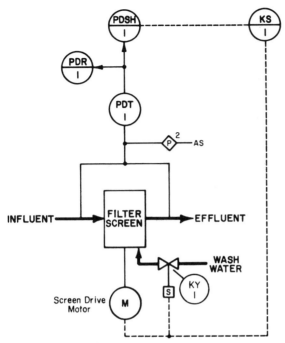

Figure 3–7. Primary treatment control system.

differential alarm contact remains closed and the screen drive motor and backwash solenoid valve remain on.

If the differential has dropped, the drive motor and solenoid will remain de-energized until the differential again builds up to the alarm point. If recording is not required, an adjustable pressure switch can be used in place of the recorder alarm to energize the auxiliary timer motor.

Sedimentation and Flotation Processes. The sedimentation and flotation processes do not require instrumentation to remove settled solids from wastewater. Instrumentation is required if pH is being adjusted for subsequent treatment needs or if flocculant chemicals are used to obtain higher separation rates.

Flocculant Addition. The addition of flocculant chemicals, which can be in the form a liquid solution or a dry chemical, is controlled through a ratio flow control system (see Figure 3–8). If the chemical is dry, some form of weight transmitter can be substituted for the chemical flow transmitter.

pH Adjustment. Adjustment of pH is achieved by a simple control loop which adds either acid or alkali, using split-range positioners on the control valves. Because of the coating action of the wastewater, installation of the pH electrode should allow for easy, periodic removal for cleaning.

Influent Metering. The influent flow can be metered by using a flow tube with a differential pressure transmitter or it can be metered by a magnetic flowmeter.

Application Data. Liquid chemical flow can usually be measured by an orifice plate. However, in some cases low flow rates may require calibrated assemblies while large flow rates may justify a magnetic flowmeter. The final control element for the chemical additive will depend on whether the additive is a liquid or a solid.

The quantity of grease in the influent can create problems for magnetic flowmeters and work is being done on heated meter bodies to minimize this problem. There is, however, a long history of successful use of flow tubes for this application.

INSTRUMENT LIST

	Tag No.	Description
Filter screen	p^2	Purge air constant differential relay
backwash control	PDT 1	Differential pressure transmitter
	PDR 1	Differential pressure recorder
	PDSH 1	Differential pressure switch–high
	KS 1	Timer and switch
	KY 1	Solenoid valve
Flocculant flow	FT 1	Flocculant flow transmitter
control	FT 2	Influent flow transmitter
	FY 1	Ratio station
	FIC 1	Flocculant flow controller
	FCV 1	Flocculant flow control valve with electro-pneumatic positioner
pH control	pHE 1	pH immersion element
	pHT 1	pH transmitter
	pHIC 1	pH indicating controller
	FCV 2	Alkali flow control valve split ranged
	FCV 3	Acid flow control valve split ranged

SECONDARY TREATMENT

Secondary waste treatment is the second major treatment in the wastewater purification process. It utilizes biochemical reactions similar to those in natural purification processes in order to remove suspended solids, settled solids, soluble organics and soluble inorganics which are present in primary treatment effluent.

In secondary treatment, the dissolved oxygen content of the wastewater is maintained at a concentration which promotes the growth of aerobic or anaerobic organisms. These organisms utilize degradable organic and inorganic matter for food. Through complex reactions, they control the concentration of soluble nondegradable

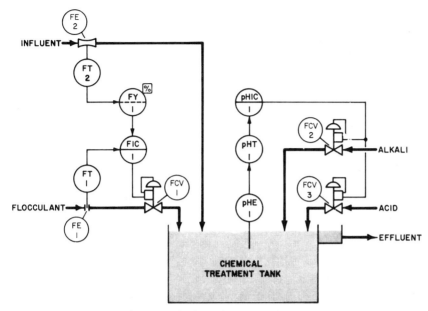

Figure 3–8. Chemical treatment.

organic and inorganic compounds containing such elements as nitrogen, phosphorus, silicon and aluminum.

Precise chemical formulation of complete biological degradation reactions is not presently possible because of two factors:

- Variations in the population of organism colonies which can determine their ability to degrade particular pollutants.

- Complex interference patterns resulting from wide variations in both concentration and composition of wastewater pollutants.

To compensate for this inability to make precise mathematical predictions of biochemical reactions, the wastewater treatment industry uses an inferential parameter, biological oxygen demand (BOD), to define the weight of dissolved oxygen utilized by organisms as they degrade or convert organic and inorganic compounds in a specific wastewater. BOD is a reasonable parameter to define the biochemical demand for oxygen to accomplish degradation of a specific wastewater, but it does not cover the amount of dissolved oxygen required for reduction of those organic and inorganic compounds which resist biological degradation, but which influence the oxygen balance of the receiving waterways. To overcome this problem, two parameters are being used to define more completely the dissolved oxygen demands of a specific wastewater. They are chemical oxygen demand (COD) and total organic carbon (TOC). Both of these

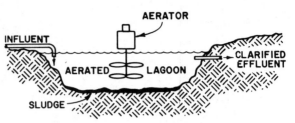

Figure 3–9. Aerated lagoon.

parameters measure total oxygen demand rather than degradation oxygen demand as measured by BOD, but, like BOD, COD and TOC do not tell us anything about the dissolved oxygen requirements of specific compounds in the wastewater.

Regardless of these limitations, BOD, COD and TOC are the best available means of defining the quantitative dissolved oxygen requirements for purification of a specific wastewater.

BOD is the most commonly used technique and is expressed in pounds of oxygen per unit of time, such as pounds per day. The magnitude of BOD loading is one significant factor in determining what type of treatment is best suited to a particular wastewater.

Treatment Process

The following three secondary treatment processes are in common use:

- Aeration
- Activated sludge
- Trickling filtration

BOD loading, available space, initial cost, and operating cost all help determine which type of treatment will be selected for a particular installation.

Aeration. Figure 3–9 illustrates an aerated lagoon tank. Here the dissolved oxygen content of the water may be controlled by any of the following methods to bring about intimate contact between air and wastewater:

- Vigorous agitation of the liquid to permit surface absorption of the air.
- Bubbling air through the liquid.
- Spraying the liquid into the air.

Through intimate contact between air and wastewater, the dissolved oxygen content of the wastewater can be maintained at a level such that the growth of aerobic organisms flourishes and biochemical degradation of pollutants is accomplished.

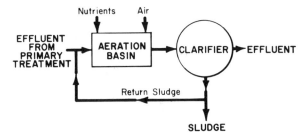

Figure 3–10. Activated sludge process.

Insoluble organic and inorganic solids precipitate to the bottom of the aeration lagoon from which they are removed as sludge. The supernatant liquid is discharged to subsequent treatment processes or to the receiving waterway. Since colonies of organisms remain in the lagoon, the removal of sludge must be handled in such a way that the colonies are not depleted to a level which is insufficient to handle the incoming load of pollutants.

Activated Sludge. Figure 3–10 illustrates an activated sludge process. This process is very similar to the aeration process. Treatment takes place in a basin where the dissolved oxygen content is maintained by agitators, compressed air, or water sprays. Here the organism colony is constantly changed instead of being continuously resident. Precipitation of insoluble organic and inorganic solids requires a clarifier. A portion of the sludge (containing organisms as well as insolubles) is constantly circulated back into the treatment tank.

Because of a short contact time in the treatment tank, nutrients are often added to accelerate organism growth and improve the effectiveness of the activated sludge process.

Trickling Filtration. In the filtration process, shown in Figure 3–11, organisms are induced to grow on the filter media (rocks or other coarse material). The raw waste trickles down over the filter media while air is blown up through the filter. After passing through the trickling filter, the wastewater then passes through a clarifier where the sludge settles out. The clarified effluent continues on either to further treatment or to the receiving waterway. Like the activated sludge process, nutrients are added to accelerate organism growth and compensate for short contact time.

Techniques which meet requirements for effluent quality are currently available to improve the effectiveness of the secondary treatment processes. The techniques involve the addition of chemical additives to improve the flocculation and precipitation of solids, nutrient additions to increase organism activity resulting in greater

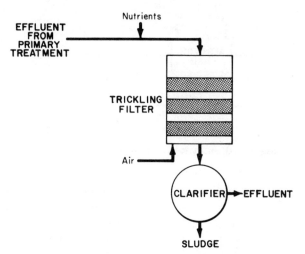

Figure 3–11. Trickling filter.

removal of pollutants, addition of pure oxygen instead of air to accomplish greater removal of pollutants, development of adaptive organisms that degrade specific pollutants and in some cases function in anaerobic conditions, and multiple stage secondary treatment involving various combinations of aeration, activated sludge and adaptive organisms.

Control System Instrumentation currently in use for secondary wastewater treatment processes can be divided into two basic categories—monitoring and control.

Since many secondary treatment plants discharge their effluent water directly into natural waterways, they can be required to monitor both flow rate and composition to verify their compliance with regulations. Usually flow rate, turbidity, and dissolved oxygen will be all that is needed, but in particular situations any of the specific ion measurements can also be required by regulating government agencies.

pH and Surge. Biochemical reactions can be adversely affected by pH and loading surges, so it may be necessary to include chemical treatment tanks or surge tanks ahead of the secondary treatment process. In the absence of surge capacity, a bypass line is often used to avoid the secondary treatment process. Instrumentation for pH control of a chemical treatment tank has been covered under the section on primary treatment.

One technique for limiting surges in the load involves controlling the level in a wet well upstream of the secondary treatment process. A typical wet-well system is shown in Figure 3–12. If the

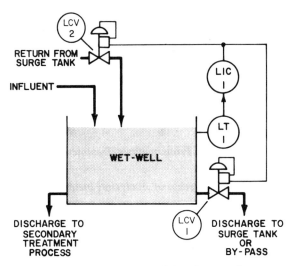

Figure 3–12. Wet-well level control.

rate of the influent flow increases and causes the level in the wet
well to rise, the level controller, LIC 1, opens the level control
valve, LCV 1, and permits some of the liquid to discharge to a surge
tank or to bypass the secondary treatment process. If the rate of
influent flow decreases and causes the level in the wet well to fall,
the level controller opens the level control valve, LCV 2, allowing
liquid from the surge tank to reenter the wet-well.

Aeration and Activated Sludge. The control systems for aeration and
activated sludge processes are quite similar to each other and can
vary from no control at all to coordinated systems such as shown
in Figure 3–13. Influent flow may be measured by means of a flow
tube, flume, or magnetic flowmeter. Because of the high flow rate
and the possible presence of grease in the influent, magnetic flowmeters require a special design of bodies and electrodes to give
accurate readings. Hence, you will find flow tubes or flumes most
frequently specified.

The ratio of chemical or nutrient additives to influence flow is
controlled by means of a ratio station, FY 1, and a flow controller,
FIC 1. Dissolved oxygen (D.O.) can vary greatly within a particular
vessel. Therefore, it is very important that the D.O. sensor be located in a position which accurately represents an average condition. A further point which must be considered in order to insure
satisfactory operation of the dissolved oxygen control loop is its
need for maintenance. Like most analytical instruments, the D.O.
analyzer is designed to measure precisely low concentrations of a
variable which is difficult to measure. To maintain its accuracy,

frequent calibration checks and cleanings of the sensing element may be necessary. Consequently, both sensor and transmitter should be placed so that the necessary work can be easily accomplished even during inclement weather.

Trickling Filter Control System. Instrumentation for the control of trickling filters can be the same as that used for aerators or activated sludge processes with the influent flow setting the nutrient feed rates and the dissolved oxygen setting the air injection rates. Figure 3–14 shows an alternative trickling filter control system.

Here, nutrient flow is controlled by a specific ion control loop which measures and controls nitrate, sulfate, sodium, or some other variable in the effluent. Accurate control of the nutrient addition can be obtained with this technique since it can compensate for load changes other than those caused by changes in the influent flow rate. Selection of the proper sensor is dependent on the type of nutrients utilized and the presence of other elements or compounds which may interfere with the measurement.

INSTRUMENT LIST

	Tag No.	Description
Wet-well level control	LT 1	Level transmitter
	LIC 1	Level controller
	LCV 1	Outlet flow control valve
	LCV 2	Inlet flow control valve with electro-pneumatic positioner
Influent flow control	FT 1	Influent flow transmitter
	FT 2	Nutrient flow transmitter
	FY 1	Flow ratio computer
	FIC 1	Nutrient flow controller
	FCV 1	Nutrient flow control valve with electro-pneumatic positioner
Dissolved oxygen control	AT 1	Dissolved oxygen transmitter
	AIC 1	Dissolved oxygen controller
	ACV 1	Air flow control valve with electro-pneumatic positioner
Specific control	AT 1	Specific ion indicating transmitter
	AIC 1	Specific ion controller
	ACV 1	Nutrient flow control valve with electro-pneumatic positioner
Dissolved oxygen control	AT 2	Dissolved oxygen transmitter
	AIC 1	Dissolved oxygen controller
	ACV 2	Air flow control valve

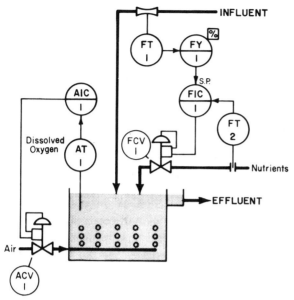

Figure 3–13. Aeration basin control.

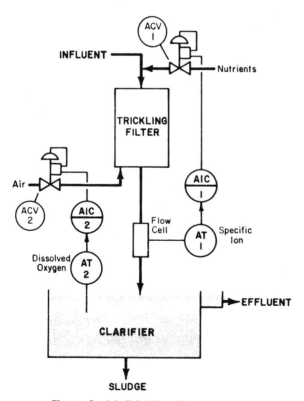

Figure 3–14. Trickling filter control.

SLUDGE CONCENTRATION

One of the major tasks in the treatment of wastewater is the removal of solids from the wastewater. Through the process of screening, flotation, or clarification, solids are accumulated and subsequently removed in a slurry form which is called *sludge*. This sludge is made up of 90 to 98 percent water and 2 to 10 percent dry solids. For economical disposal, sludge should contain at least 25 percent dry solids. Concentration and dewatering processes must be utilized to treat sludge withdrawn from the primary or secondary waste treatment processes.

Processes used to concentrate sludge generally require little mechanical energy and can produce sludge of 10 to 20 percent dry solids content. Digestion, clarification and flotation are the most common concentration processes currently in use in the wastewater treatment industry.

DIGESTION
Sludge digestion involves the biological decomposition of its organic content under either aerobic or anaerobic conditions. It may take place in a covered vessel called a *digester* or in a tank which is called a *digestion tank*. Heat is often added to maintain sludge temperatures of 90 to 95°F (30 to 35°C) to shorten biological reaction time.

As biological action takes place, the volatile content of the sludge is reduced by converting some organic compounds to gas and others to liquids. This permits liquid and solid separation, giving a higher concentration of solids in the sludge and, at the same time, reducing the sludge's obnoxious odor. Gases, given off during digestion, are generally used as combustible fuels to heat or incinerate sludge.

CLARIFICATION
By definition, clarification is any process whose primary purpose is the reduction of a concentration of matter suspended in liquid. Strict adherence to this definition would include screening, digesting, sedimentation, flotation and other processes. In actual practice a clarifier is a tank in which gravity is utilized to remove solids from the wastewater. These tanks are usually circular in shape and use mechanical scraper systems for removing settled solids without disturbing the overlying clarified liquid.

As used to concentrate sludge, clarifiers have one major problem—the long time period required to clarify a high solids content slurry. This problem has led to the use of reaction-type clarifiers, in which chemical additives are used to improve the precipitation of solids through both chemical reactions and the filtering effect of the resultant floc as it settles.

A further modification of this technique utilizes chemical precipitation but feeds the clarifier from the bottom so the incoming liquid passes up through the precipitated solids, or floc, which acts as a filter bed. The clarified liquid is drawn off near the top of the clarifier. This type unit is called a solids-contact clarifier.

Prior to the development of reaction-type and solids-contact clarifiers, the flotation process was a key element in the concentration of sludge for dewatering and final disposal. Using smaller air bubbles to aglomerate and float solids suspended in the liquid, it is possible to handle effectively high concentrations of solids in the floating scum which is skimmed off the flotation tank surface.

FLOTATION

Sludge which has passed through the concentrating processes will have a solids content of 15 to 25 percent depending on how it has been treated and the types of solids it contains. To be suitable for final disposal it must be converted from a liquid to a semisolid form, under 75 percent moisture and readily movable with a shovel.

SLUDGE DEWATERING

Many processes are available to accomplish this transition, including:

- Draining
- Filtration
- Vacuum filtration
- Centrifugation

Each process has both advantages and disadvantages which must be considered in selecting the process best suited to a particular type of sludge.

When the sludge composition is such that it can easily be dewatered, simple gravity filtration procedures effectively dewater it for final disposal.

Gravity Filtration Sludge Dewatering

Figure 3–15 shows a gravity filter which uses a fine mesh wire filter medium. The wire mesh rotates continuously over the dewatering cell where the incoming sludge is discharged. The major portion of the water content passes through the screen leaving behind the solid matter. This layer of solids is formed into a rolling cake of solids with a low moisture content. Any excess cake is discharged to a conveyor which removes it for final disposal. If a sludge has a high concentration of colloidal solids, they will limit the use of this type unit since they will pass through the screen mesh filter medium. Sludge containing high percentages of slime will also make this type of filter ineffective since it will retard formation of the rolling cake, blinding the filter medium so it will not allow water to pass through the dewatering cell.

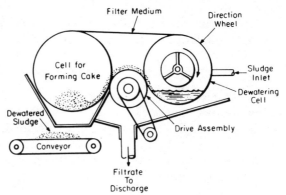

Figure 3–15. Sludge dewatering by filtration.

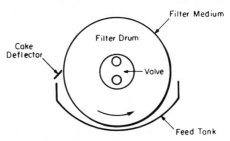

Figure 3–16. Drum vacuum filter.

Vacuum Filtration Sludge Dewatering

When sludge composition presents problems which gravity flotation cannot handle, vacuum filters are often used. Two basic types of vacuum sludge filters commonly encountered are:

• The drum filter

• The continuous belt filter

Drum Filter. In a drum type unit (see Figure 3–16), sludge is drawn against a revolving drum, which is submerged in the slurry tank, by a vacuum inside the drum. The surface of the drum is a fine mesh screen, cloth, or fine-wound coil springs. On this surface the major portion of the solids are retained. The water and some fine solids pass through the surface into the drum and into the effluent discharge. As the drum rotates, the layer of solids coating the surface, called a filter cake, is rotated out of the slurry tank and air dried as it continues on for removal by scrapers, flexing, or spray backwashing. After the cake has been removed, the filter medium must be backwashed to remove residual solids which can blind the filter medium and limit its ability to act continuously as a filter.

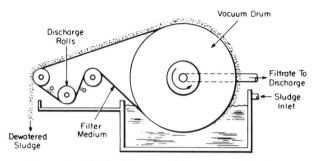

Figure 3–17. Continuous belt vacuum filtration.

This unit has a higher filtration ability per square foot of surface than a gravity unit and can successfully handle sludge containing limited amounts of slime. If, however, the sludge contains over 50 percent of less than 200 mesh particles, the vacuum filter is limited in its ability to build up a suitable cake. In other words, the cake removal rate is so low that the filter is ineffective.

Continuous Belt Filter. A continuous belt vacuum drum filter, shown in Figure 3–17, can overcome most of the difficulties which limit the effectiveness of a drum vacuum filter. This unit forms the filter cake on a continuous belt filter medium which passes over small diameter discharge rolls, which can break thin cakes as well as thick cakes. Backwash sprays are used to remove any residue from both sides of the filter medium so that blinding of the medium does not occur.

Since the backwash can be isolated from the feed tank, chemicals can be utilized to improve the cleaning of the filter medium during backwash. Continuous belt vacuum filter units can handle sludge feeds containing 100 percent of less than 200 solids. They also lend themselves to precoating techniques which can produce filter effluent with as little as 10 ppm solids content.

Centrifugal Sludge Dewatering

Centrifuges have long been in use for dewatering sludge. Recent improvements in design have greatly improved their ability to give high clarity effluent without appreciable loss in solids dewatering ability. By controlling the speed of the machine, wide variations in feed rates can be handled without loss of efficiency.

More than with any other dewatering process, centrifuge performance is subject to the handled waste characteristics. It is also highly subject to feed concentration. A 5 percent increase in solids content in the feed can result in a 250 percent increase in solids discharge without reduction in recovery efficiency.

Other Sludge Dewatering Techniques

In addition to the methods of sludge dewatering described above, you will occasionally find cases where evaporation or pressing is used. Evaporators use heat and vacuum to remove water and presses use rolls, where pressure and spacing separate solids from the water. Flotation may also be classified as a dewatering process since it can produce a sludge of 25 percent dry solids content.

SLUDGE DISPOSAL

Once sludge has been concentrated and dewatered to a spadable consistency, it is ready for final disposal by one of three methods:

- Landfill
- Multiple hearth incineration
- Fluidized bed incineration

LANDFILL

While landfill was the common method of disposal for many years, lack of space, ecological problems resulting from surface water run-off, and the rapidly increasing quantities of sludge have necessitated the termination of this method of disposal in most areas. Incineration resolves these problems and has become the most common method of sludge disposal. Wet oxidation and steam injection processes are also acceptable methods of sludge disposal where specific conditions economically justify their use.

MULTIPLE HEARTH INCINERATION

Multiple hearth incineration (see Figure 3–18) is adaptable to wide variations in sludge volume and composition. In well-run, well-designed installations, sludge can be reduced from 60 to 75 percent solids content to $\frac{1}{10}$ of its original weight without requiring supplemental heat addition.

Figure 3–18 shows a five hearth unit consisting of a refractory lined shell, refractory hearths and a rotating air-cooled center shaft with air-cooled rabble arms and teeth. Sludge is fed through the roof of the unit by a starfeeder, screw conveyor or other heat sealing method. Once it reaches the top hearth, the sludge is moved across the hearth by the rabble arms until it drops through drop holes to the next hearth where the process is repeated. Finally the incinerated sludge reaches the bottom of the incinerator.

In the top hearths the sludge is dried in a drying zone to less than 50 percent water content. Following the drying zone is an incineration zone where temperature is maintained at about 1600°F (870°C) and the sludge is converted to ash.

A cooling zone follows. Here incoming air is utilized to cool the ash before discharge. At the same time, the air is heated as an aid to sludge combustion in the incineration zone. Grease, if it has

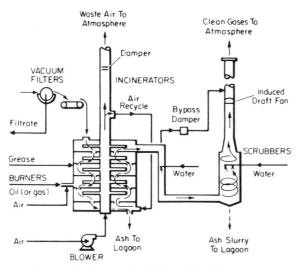

Figure 3–18. Multiple hearth incinerator.

been separated from the sludge, is injected directly into the incineration zone since it does not require drying before combustion. While sludge of 25 to 40 percent solids content and 70 percent volatile content in the solids can sustain its own combustion without supplemental fuel, a heat source is necessary to bring the furnace and sludge content up to a temperature where combustion can initiate a self-sustaining operation.

A portion of the air used to cool the rotating center shaft and rakes is recycled back to the bottom hearth. It acts as the coolant for the hot sludge ash here before passing into the incineration zone and the drying zones. Then it is exhausted at approximately 500 to 1000°F (260 to 535°C). The exhaust gases are cooled and cleaned by water spray scrubbing before their discharge into the atmosphere.

Ash from this type of unit is sterile, inert and $\frac{1}{10}$ of the initial sludge volume. It can be handled mechanically, pneumatically, or hydraulically for removal to a landfill or other uses. Exhaust gases from the scrubber are sufficiently clean to meet air pollution standards.

Fluidized bed incinerators can compete with multiple hearth incinerators in cost and performance. Figure 3–19 shows a fluidized bed incinerator. Evaporation of moisture and combustion of organics takes place simultaneously under oxidizing conditions at approximately 1500°F (815°C) in the bed.

Hot gas discharged from the reactor at about 1500°F (815°C) is passed through a preheater to heat the combustion air to about

FLUIDIZED BED INCINERATION

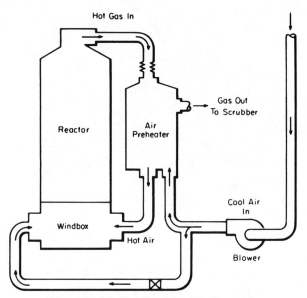

Figure 3–19. Fluidized bed incinerator.

1000°F (535°C) before it is delivered to the reactor windbox. To start the process, external heat must be added either to the air preheater or the windbox to initiate combustion. On sludges of 30 percent or less dry solids, additional heat will be required to sustain the incineration reaction.

4 Petroleum

OIL AND GAS PRODUCTION

Petroleum hydrocarbons can be classified into three basic groups: gas, condensate, and crude oil.

CLASSIFYING HYDROCARBONS

- Gas contains mainly methane with some ethane and propane.

- Condensate contains mainly propane, butane and pentane with some hydrocarbons as heavy as nonane present.

- Crude oil contains all petroleum hydrocarbons from hexane throught asphalt.

These three groups usually exist in some combination rather than independently.

If the gas contains mostly methane with no recoverable quantities of ethane or propane, the gas is called *dry gas*. When the quantity of heavier hydrocarbons in the gas reaches an economically recoverable level, the gas is called *wet gas*. Wet gas formations usually contain some degree of condensates which can either exit in small quantities with the gas or be recovered from the formation in a liquid form. In some cases the amount of condensate in a gas well can be so great that the gas is dissolved in the condensate at the formation pressure.

Wet gas can also be produced from a formation containing crude oil. In this case, the gas is called *associated gas* whereas the dry gas and wet gas previously mentioned are nonassociated gases. Associated gas comes from a gas cap over the oil in the formation. If no gas cap exists over the oil, gas may still be produced by the well because the gas comes out of an oil solution as the oil is produced. This gas is called *dissolved gas*. Crude oil can occur without any appreciable dissolved gas in some formations.

Produced gas and oil also contain other materials. The most common of these are water, carbon dioxide, nitrogen, helium and hydrogen sulfide. Produced gas is a major source of helium and sulfur. Water and carbon dioxide have no real market value but must be produced in order to obtain the hydrocarbons. They are often used to aid in the recovery of more condensate or crude oil from partially depleted formations.

OIL AND GAS TRAPS

Oil and gas are found in porous rock formations called *traps*. This can be sand or some conglomerate which actually supports the structure above the porous formation. The space between the grains of porous material provides a path through which water, oil and gas can flow. The surrounding formation is of some nonporous material, such as shale or bedrock. As shown in Figure 4–1, water flowing through the formation carries the oil and gas with it until a high point in the formation is reached. Since oil and gas are lighter than water, they will accumulate in this high area until the space between the bottom of the oil and the bottom of the formation is only large enough to provide space for the water to continue flowing.

The formation structure will frequently shift, as shown in Figure 4–2; these shifts, called *faults*, will interrupt the formation structure. If two or more faults occur in the same formation, the traps can become isolated from the porous formation. This is only one of many ways in which isolated traps can occur.

When subsurface water is flowing, the formation pressure is the hydrostatic pressure of the water itself. This pressure is proportional to the hydraulic gradient between the water source and sea level. If the trap becomes isolated from the porous formation, changes can occur which causes the trap pressure to increase or decrease within the strength limits of the particular formation.

Discovery pressures in hydrocarbon-bearing formations generally vary from approximately 1 000 psi (6 900 kPa) for shallow formations to over 17 000 psi (117 000 kPa) in formations as deep as 20 000 feet (6 100 m). Pressures as low as 100 psi (690 kPa) and as high as 30 000 psi (207 000 kPa) are occasionally found in some formations. Usually the formation pressure is sufficient to drive the gas or oil to the surface when a well taps the formation. This pressure can decrease as gas or oil production continues.

PRODUCTION

Drilling

To remove gas or oil from a producing formation, the formation is tapped by one or more wells drilled through the surface rock structure. After the well is drilled, a steel casing is placed in the hole to maintain the hole integrity.

To prevent seepage between the casing and the well bore and to help support the casing and the rock structure, cement is forced between the casing and the well bore. If the well is deep or passes through formations which must be supported or blocked off, a portion of the casing can be set in place and drilling can continue. In this instance, the casing for the lower section of the well is smaller

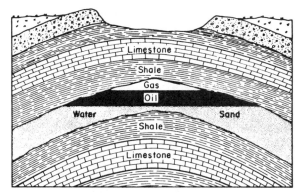

Figure 4–1. Anticline or dome with trap for oil and gas.

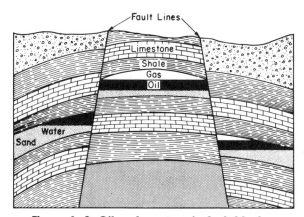

Figure 4–2. Oil and gas trap in fault blocks.

in diameter than the upper casing to permit it to pass through the upper section. This is called a telescoped casing and can occur several times in deep wells (see Figure 4–3).

The well can pass through several producing formations before the primary target is reached. The depth and pressure of each formation is logged during the drilling in case a sufficiently productive formation other than the target one is discovered, which would warrant production from more than one formation in the same well. This is called a *multiple completion*. Occasionally the production rate from a well is not sufficient to justify further development and it is abandoned. This is called a *dry hole*.

Once a well is drilled to the expected producing formation and the casing set, the portion of the casing inserted is perforated to permit flow into the casing from the greatest possible area (see Figure 4–3. If the flow rate into the well is not as great as expected, various

Well Casing and Piping

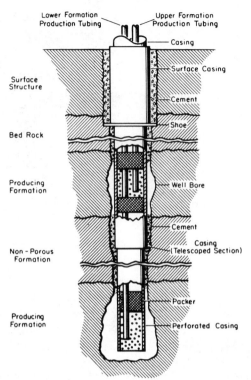

Figure 4–3. Typical well structure showing a dual completion well.

steps can be taken to improve the porosity of the surrounding rock formation.

Production flow from a well is usually up through the production tubing, not through the casing. There are two basic reasons for this. The first is that the casing at the surface may not be strong enough to take the full formation pressure and, secondly, a smaller tube has a greater flow velocity. This velocity carries any entrained solids or gases out of the tube with less separation than the larger tube. The casing above the producing area is usually sealed off with a packer so that there is less dead space above the producing area. Formation pressure is not applied to the upper casing. In multiple completions, there are packers above and below the upper producing areas.

Each producing formation of a multiple completion well is produced separately because of the usual differences in composition and pressures. In this case, a separate tube is run down the well to each producing area (see Figure 4–3). Care must be exercised in producing oil wells which have gas or water drives. Excessive production rates can cause channeling of the water or gas to the well. This can cut off the flow of oil, resulting perhaps in an abandonment of the well.

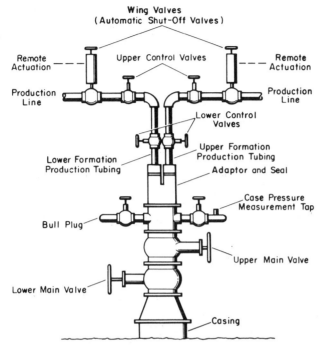

Figure 4–4. Typical well head arrangement for a dual completion well.

Since extremely high pressures exist in the well, considerable care is taken to close in the well with as little as possible of the well production tubing and casing exposed above the ground at the well head. Usually two manual and one automatic shutoff valves are close coupled to the top of the tubing, as shown in Figure 4–4. The automatic valve, called a *wing valve*, is tripped by a switch in the downstream tubing to protect the downstream equipment. Energy to operate this valve is taken from the downstream tubing and stored to assure reliable operation of the valve. If remote monitoring of a well is used, the position of the wing valve is usually included. Control systems will be reviewed in the individual sections which cover, in detail, the various methods of well production.

NATURAL LIFT WELLS

When the pressure in a crude oil or condensate producing formation is greater than the hydrostatic head of the crude oil or condensate in the production tubing between the formation and the surface, the liquids will flow from the well. This is called *natural lift* and it is a primary form of recovery.

If the formation is pressurized by flowing water under the trap, the pressure at the well head is equal to the hydraulic gradient pressure of the water minus the hydrostatic head of the oil in the tubing. As oil flows from the well, the well head pressure is reduced by friction in the tubing, resistance to oil flow through the formation, and resistance of water flow into the trap. When the flow from the well is stopped, the well head pressure will return to the same pressure as before the flow started.

A gas dome over the oil may also furnish the drive pressure for oil production. If gas is the only source of drive pressure, the formation pressure will decrease as oil is produced due to the expanded gas volume. The production of gas alone from such a formation will make it more difficult to recover the oil.

When both gas and water act as the oil drive pressure, gas may be produced from the formation. However, if gas is removed indiscriminately, the oil well may start producing water and the gas well may produce oil.

Some oil traps do not contain any natural drive. Although the discovery pressure of the formation may be very high, the drive pressure drops rapidly with little or no production. When this occurs, other means of producing this particular type of formation must then be used. For example, it is frequently possible to add gas or water to the formation to produce oil in the same manner as natural drive.

There is a rate at which a well may be worked for maximum liquid recovery in the formation. Since naturally occurring hydrocarbons are a finite government valuable natural resource, regulations have been established to prevent premature formation depletion. Part of these regulations outline allowable limits to the maximum flow rate of the well based on periodic tests to determine the characteristics of the well and the formation. Other regulations establish a maximum allowable flow rate for maximum recovery in an attempt to utilize all available sources equally. This is called the allowables for a well and may be changed depending on demand and economics. The rate at which a well is actually produced is a function of demand up to the allowable flow rate.

In most cases, the crude oil or condensate is individually piped to a central location called a *battery*. At the battery, the well flow is metered and the water and dissolved gas is removed. If the liquid contains a high percentage of gas or water, it may be separated at the well site before the oil goes to the battery.

CONTROL SYSTEM

Liquid flow rates from a producing well are usually controlled by means of a fixed restriction, called a *choke*, in the production line at the well head. The choice of choke used is based on its ability to obtain the desired flow rate with all of the downstream equipment in operation. Allowables for a well are based on a monthly volume. To meet this requirement, the well flows until the allowable volume is reached and is then shut down for the balance of the month. If a different flow rate is required, the choke can be changed. The choke size can also be changed if the well plugs up or the formation pressure decreases. If frequent choke size changes are needed, an adjustable choke can be used. This device provides a selection of several different choke sizes in a single unit. The adjustable choke can be automated to allow the choke size to be changed from a remote location (see Figure 4–5). When a remote monitoring system is used, the choke position is usually indicated.

When the well flow reaches the monthly allowable limit, the well is shut-in. If the well is remotely monitored, a shut-in valve can be remotely activated when the allowable flow volume has been reached.

Well head pressure, PI 1, and casing pressure, PIV 2, are indicated at the well site. When remote monitoring is used, these values can be indicated as either a full value or a deviation from a normal value. Occasionally, when remote monitoring systems are used, it is advisable to shut in the well automatically when the well head pressure decreases as an indication of well plugging. The pressure controller, PC 1, is a manually reset trip type controller which closes the shut-in valve through solenoid valve SV 1 when the well head pressure is below a preset value. The energy used to operate the shut-in valve may be electrical or gas, depending on availability.

A second reset trip type controller, PC 2, protects the downstream equipment from high pressure by closing the wing valve. This controller usually operates on energy stored at the site since its safe operation is critical.

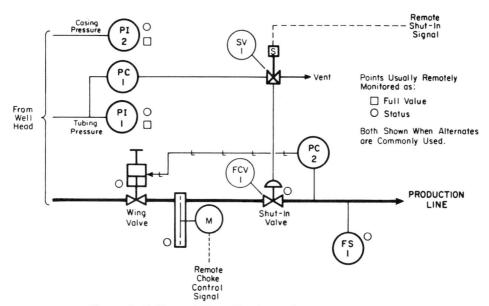

Figure 4–5. Natural lift well—Control system.

When the well site is remotely monitored, the positions of the shut-in valve and the wing valve are usually monitored. As a double check on the well and valve operation, a flow switch, FS 1, is also often used for remote indication.

APPLICATION DATA

Devices in contact with crude oil flow have a great tendency to plug up with paraffin and sand. Therefore, all valves, gauges, pressure controllers and flow switches must be designed to operate satisfactorily under these conditions.

Energy for the operation of automatic devices at the well site is a problem because of the remote location of most wells. Dissolved gas from the crude oil can be used in some cases. Care must be taken in equipment selection with using gas since it can be corrosive or contain condensate.

MECHANICAL LIFT WELLS

Mechanical lift wells are used to recover crude oil from a producing formation when other means of recovery are not practical. This means does not require pressure in the formation to remove the oil. The main requirement is that the oil flow to the well bore must be in sufficient quantities to make the process economical.

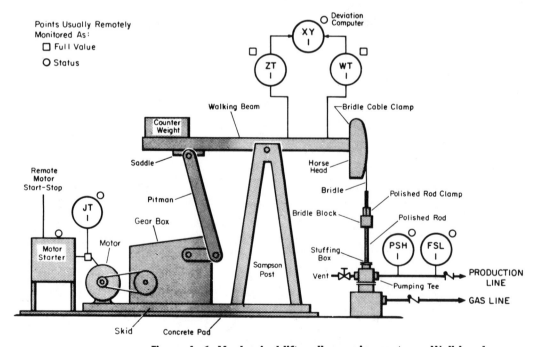

Figure 4–6. Mechanical lift well pumping system—Well head.

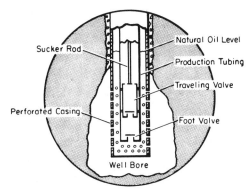

Figure 4–7. Mechanical lift well pumping system—Well bore.

The mechanical lift well, shown in Figure 4–6, operates on the same principle as a piston pump. The production tubing is the cylinder, placed well below the natural oil level in the formation. A check or foot valve at the lower end of the production tubing acts as the pump suction valve. The pump discharge valve is a check valve located in the piston. The piston and its valve are called the *traveling valve* (see Figure 4–7).

Both valves permit only unidirectional oil flow. As the traveling valve moves upward, its inner valve is closed and the entire column of oil above it is lifted toward the surface. The upward motion of the traveling valve reduces the pressure under it, causing oil to flow into the production tubing through the foot valve.

VALVE OPERATION

When the traveling valve descends, the foot valve closes and the traveling valve is forced open. The full weight of the column of liquid is supported by the foot valve in this part of the cycle. As the cycle is repeated, oil is produced and flows into the production tubing with each upward stroke of the pump. When the oil reaches ground level, it flows through a pumping tee into the production line.

The traveling valve is lifted by a shaft inside the production tubing, called a *sucker rod*. On the upward stroke, this rod lifts the total column of liquid. Since this type of pump is used on wells from a few hundred feet to more than 15 000 feet (4 500 m) deep, the oscillating strain on the sucker rod can be very great. This causes the rod to stretch on the upward stroke. Therefore, the length of stroke of the traveling valve is less than the length of the stroke of the sucker rod at the surface. On the downward stroke of the traveling valve, the full weight of the liquid column is supported by the production tubing. Free vertical movement of the lower end of the production tubing due to this oscillating strain reduces the displacement of the pump on the downward stroke. Therefore, the

PISTON AND SUCKER ROD

tubing is usually anchored to the casing to reduce this motion. Guides are used between the casing and the tubing, to prevent buckling of the tubing, and between the sucker rod and the tubing to reduce wear.

The sucker rod emerges from the well at the pumping tee through a stuffing box which prevents leakage of the oil. To reduce the friction and improve the seal at the stuffing box, the sucker rod is machined to a uniform smooth surface at this point. This section of the sucker rod is called a *polished rod*.

PUMP ACTUATION

The pump actuating device is usually purchased as a skid mounted unit, placed on a concrete pad for stability. Electric motors are often used as the source of mechanical energy with a gear box reducing the motor speed. The connecting linkage between the gear box and the walking beam converts the rotary motion of the gear box to a vertical oscillating motion. This connecting linkage permits some adjustments of the pump stroke amplitude. The walking beam pivots about the top of the Sampson post. Cables connect the sucker rod to the horsehead located at the end of the walking beam. To keep the lifting force in line with the sucker rod throughout the stroke, the horsehead is an arch with a radius to the pivot point of the walking beam. A counter weight on the walking beam keeps the load on the motor and gear box uniform during the pump cycle. Stroking speeds up to about 20 cycles per minute are used.

GAS AND CRUDE OIL

As the crude oil flows into the well casing, dissolved gas may separate from the liquid. Since a buildup of gas pressure at this point will reduce the flow rate of oil into the well, the gas is usually taken from the well side by a low pressure gas gathering system.

Crude oil from the well is usually piped directly to a central battery for metering and treating before sale. Older installations and a few unusual new installations treat the oil at the well site and store it in tanks there for sale.

CONTROL SYSTEM

Motor Control

Mechanical lift wells can be operated from a remote location. Here the motor starter is remotely actuated and the status of the motor starter is monitored.

The gear box can be equipped with a brake mechanism to stop the pump at a part of the cycle where it can be most easily restarted. Adjustment of the pumping rate is possible by changing the belt pulleys which connect the motor to the gear box.

Pump Monitoring

Pump operation, motor load, and flow are monitored by two devices. The transmitting wattmeter, JT 1, attached to the drive motor, measures the motor load. The flow switch, FSL 1, in the production line goes into alarm condition on low flow.

Since the pump is a positive displacement device, it can produce very high well head pressures. The pressure switch, PSH 1, in the production line is used to stop the pump when the set pressure is exceeded. The status of this switch may also be remotely indicated.

Pump Stress. A mechanical lift pump is subjected to many types of stresses in addition to normal wear. Since the main part of the pump is in the well, the exact cause of a malfunction must be determined indirectly by measuring the position of the piston and the force on the piston. The shape of the data plotted on a graph, called a *dynamometer card*, can usually indicate the cause of a malfunction. Dynamometer cards are made periodically for each well to check for problems. Common malfunctions are valve leakage, valve sticking, excess piston or sucker rod friction, loose tubing anchor or a pump-off condition due to an excessive pumping rate.

When a remote monitoring system is used, it may be desirable to monitor continuously the dynamometer data. This is done by measuring the position of the walking beam with a motion transmitter, ZT 1, and a load transmitter, WT 1. From simultaneous readings of these two valves, a dynamometer card is plotted at the remote. This is compared at the well site by an analog computer, XY 1, to a normal dynamometer card. The deviation for the well from the normal card is indicated at the monitoring station.

Application Data

Usually, the flow rate of the gas from the well is so low that it is not metered except as a total flow from the gathering system. Since most mechanical lift wells are driven electrically, power is available to operate electronic devices. The well stops running when the power fails. Therefore, no power backup for the control system is required. Electrical power at the well site is usually not of good quality because of the motors on the line and an increased possibility of lightning effects. Therefore, a separate well regulated power supply may be required for instruments used at a well.

Crude oil can contain paraffin which coats everything it comes into contact with. Because of this, any devices in contact with the oil must be immune to this coating action.

The equipment at a well site is large and rugged. Control equipment located at the well should be rugged enough to withstand accidental abuse without damage.

THE CENTRAL BATTERY

When there are a large group of wells in a producing area operated by a single company, it is frequently more economical and efficient to perform the separation and measurement of the well effluent at a centrally located site rather than at each well. This site is called a *central battery*. In this type of operation, the effluent from all of the wells is run through a single large separator called the *production separator*. To check the production rate and operation of each well, the effluent of a single well is periodically diverted to a second smaller separator called the *test separator*. The measurement of the gas, oil and water flow volumes while a well is connected to the test separator is used to prorate the total flow volume of the central battery from each well.

If large volumes of gas are produced by a well, the gas, oil and some of the water can be separated at the well site. The gas and oil are then piped individually to the central battery, as shown in Figure 4–8. To check the gas flow volume of each well as a prorated portion of the total central battery gas flow volume, the gas from a well is passed through a test meter at the same time as the oil from that well is passed through the test separator.

Water produced from a well is not usually separated from the oil at the well site when a central battery is used because extra handling is required. If it is separated, only the free water is removed and emulsion-breaking chemicals are not injected into the well site separator unless it is necessary to ease the piping of the oil to the central battery.

Oil leaving the central battery is usually directly piped to a customer's tank or pipeline. The specifications of the oil (which changes ownership at this point) generally allow a small amount of water.

Gas leaving the central battery may also be sold, going into a customer's gas-gathering system. If the gathering system pressure is high, the gas may have to pass through a compressor as it leaves the central battery. Usually the only requirement of this gas is that all liquids be removed. Therefore, the gas is passed through a separator before it leaves the central battery. If the gas is sour (i.e., contains corrosive contaminants), a gas treating unit is used at the central battery to remove these contaminants before the gas is accepted into the gathering system.

CONTROL SYSTEM

As shown in Figure 4–8, the oil flow is from the well through the oil production manifold to the production separator. Gas flows through the gas production manifold to the output gas line. A cycle controller, KJC 1, actuates the three-way valves in the lines from the wells. This diverts the flow from the production to the test manifolds for the tested well.

Oil Flow

The oil from the test separator is measured by the positive displacement flowmeter, FQI 6. Since the water content of this oil cannot be measured as oil, the flowmeter's output is corrected by the net oil computer, FY 6, using a capacitance probe, AE 6, to

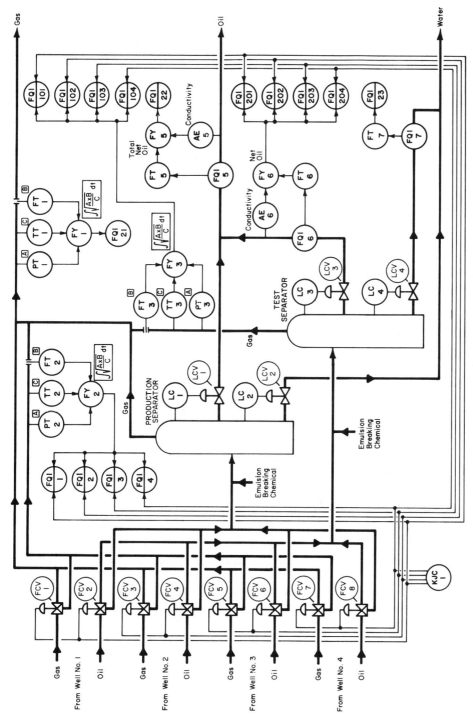

Figure 4–8. Typical central battery control system.

111

measure the water content of the oil. The output of the net oil computer, FY 6, is a pulse per unit volume of oil. These pulses, read directly as barrels of oil, are counted on a digital counter, FQI 2XX. When a particular well is on test, the cycle controller, KJC 1, allows only the counter associated with that well to receive the pulses from the computer.

Gas Flow

In a similar manner, the gas flow from the test separator is measured by means of an orifice plate. The flow volume in standard cubic feet is computed by the gas flow computer, FY 3. The computer output is a pulse representing 1,000 standard cubic feet (m^3) of gas. The pulses, read directly as standard cubic feet (m^3), are counted on a digital counter, FQI 1XX. Only that counter associated with the well on test, as determined by a signal from the cycle controller, counts the computer pulses.

Manifold Purging

At the beginning of each test cycle, the test manifold must be purged and the test separator given time to adjust to the new equilibrium condition before the test flow measurements can be made. Therefore, the cycle controller, KJC 1, does not signal the flow indicators, FQI 1XX and FQI 2XX, to count until after the well has been connected to the test separator for some time.

Water Flow

Water flow from the test separator or total water flow can be measured and read periodically to determine the overall oil-to-water ratio. A large departure from the normal ratio may indicate a problem at one of the wells. Then, a more detailed test can be made to determine which well and what corrective action should be taken. A positive displacement flowmeter, FQI 7, and a pulse counter, FQI 23, can be used for this purpose. If measurement of the water produced from each well is required during the test cycle, this flowmeter can be connected to a pulse counter for each well in the same manner as the oil flow indicator, FQI 2XX.

Well Gas

Gas from the tested well passes through the gas test manifold to a flowmeter. This flow is measured by an orifice plate. The flow volume in 1000 standard cubic feet (cubic meters) is computed by the gas flow computer, FY 2. The pulse output of this computer is indicated on the pulse counter, FQI X, in a manner similar to the pulse counter FQI 1XX.

Total Gas Flow

Total gas flow from the central battery is also measured by an orifice plate and the flow in 1000 standard cubic feet (m^3) computed by the gas flow computer, FY 1. This computer value is digitally indicated on the pulse counter, FQI 21.

Total Oil Flow

Total oil flow is measured by the positive displacement meter, FQI 5, and corrected for water content by the net oil computer, FY 5.

This computer uses a capacitance probe to determine the percentage of water in the oil. The flow volume is indicated on the pulse counter, FQI 22.

Remote Monitoring

The well cycle controller, KJC 1, selects the wells to be tested on a regular time-based cycle. When a remote monitoring system is used, the tested well is remotely indicated. In some cases, the well test cycle is entirely determined by the remote monitoring system. A combination of these systems is also used to permit either time cycle testing or selected testing by the remote monitoring system.

Even though a remote monitoring system is used to read all flow indicators, local digital indicators are used as data protection against transmission loss. As further protection against data loss, the direct digital indicators on the total oil flow positive displacement meter, FQI 5, and the total gas flow temperature, pressure and orifice plate differential pressure are recorded. This also enables production to continue with a power loss.

Emulsion-Breaking Chemicals

The addition of emulsion-breaking chemicals to the separators is usually fixed at a small rate by means of a needle valve and rotameter. Since the chemical volume is small, the use of a ratio metering system is not justified. The chemical flow to the test separator can be interrupted by the cycle controller if there is no well on test.

Application Data

In most cases, power is available at a central battery although it can be subject to interruption because of the remote location. Therefore, the system must be designed to continue operation without loss of data and to restart automatically without the need of manual readjustment.

Simple local control, such as the separator level controls, are either gas powered or mechanically operated for reliability. Since the gas can sour, the materials used in the construction of these controllers must be carefully considered.

INSTRUMENT LIST

	Tag No.	Description
Battery feed valves	FCV 1–8	Flow diversion valve
Total gas flow computing loop	PT 1	Gas pressure transmitter
	TT 1	Gas temperature transmitter
	FT 1	Gas flow transmitter
	FY 1	Total gas flow computer
	FQI 21	Totalizer-pulse counter

	Tag No.	*Description*
Well gas flow computing loop	PT 2	Gas pressure transmitter
	TT 2	Gas temperature transmitter
	FT 2	Gas flow transmitter
	FQI 1–4	Totalizer-pulse counter
Test separator	LC 3	Oil level controller
Oil level control loop	LCV 3	Oil flow control valve
Water level control loop	LC 4	Water level controller
	LCV 4	Water flow control valve
Test gas flow computing loop	PT 3	Gas pressure transmitter
	TT 3	Gas temperature transmitter
	FT 3	Gas flow transmitter
	FY 3	Test gas flow computer
	FQI 101–104	Totalizer-pulse counter
Production separator	LC 1	Oil level controller
Oil level control loop	LCV 1	Oil flow control valve
Water level control loop	LC 2	Water level controller
	LCV 2	Water flow control valve
Total net oil computing loop	FQI 5	Oil flowmeter
	FT 5	Oil flow transmitter
	AE 5	Conductivity probe
	FY 5	Total net oil computer
	FQI 22	Totalizer-pulse counter
Net oil computing loop	FQI 5	Oil flowmeter
	FT 6	Oil flow transmitter
	AE 6	Conductivity probe
	FY 6	Net oil computer
	FQI 201–204	Totalizer-pulse counter
Water flow measuring loop	FQI 7	Water flowmeter
	FT 7	Water flow transmitter
	FQI 23	Totalizer-pulse counter

LEASE AUTOMATIC CUSTODY TRANSFER UNIT

Oil leaving a central battery is piped directly to a large storage tank for batch shipment through a pipeline to the refinery. The refining company is not the same company as the oil producing company. The oil transfers ownership from the producing company to the refining company, called *custody transfer*, as it leaves the central battery. The refining company receives oil from several different batteries or producing companies in the same general area.

The oil is purchased by the refining company on the basis of volume delivered at a maximum water content. Usually, this water content cannot be greater than one percent. Since no premium is paid for oil with less than the specified water content, it is to the advantage of the producing company not to remove more water than necessary from the oil.

Since it is difficult to control the water content of the oil leaving the separators at the central battery, the oil is passed through a process unit called a lease automatic custody transfer unit (LACT) which is specifically designed to control the amount of water in the oil. This unit contains a volume flowmeter to measure accurately the delivered volume of oil. The accuracy and operation of the LACT unit water content control system and flowmeter is periodically checked by both the producing and refining companies since payment is made based on these values.

The water in the oil supplied to the LACT unit is mainly in an emulsified form. The emulsion is most easily broken down for separation by the application of heat. This is usally done in a water bath heater (see Figure 4–9), rather than a direct-fired heater for reasons of safety and convenience. If temperatures are needed which are higher than possible using water at atmospheric pressure, heating mediums other than water can be used.

After the emulsion has been heated, the water and oil form two phases and can be separated in a vented separator. The water from the separator is stored in a tank to await pickup by a tank truck. Oil from the separator flows into a tank which acts as a surge tank before being piped to the large storage tank. As the oil leaves the surge tank, its water content is checked. If it is below the specified level, the oil flows to the storage tank. When the maximum allowable

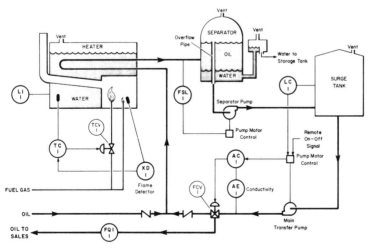

Figure 4–9. Lease automatic custody transfer unit (LACT).

water content is exceeded, the oil is recycled through the LACT unit for additional processing.

CONTROL SYSTEM

The temperature of the oil from the oil heater is not directly controlled since it is not critical. The water bath temperature is controlled by an on/off temperature controller, TIC 1, which controls the fuel to the main burner.

A pilot flame detector, XD 1, prevents operation of the main fuel valve, TCV 1, when there is no pilot flame. Oil passing through the tubes in the water bath closely approaches the water bath temperature. The oil temperature is controlled in this manner.

Since the oil-water interface level in the separator is usually controlled by a water leg type of overflow system and the oil level is controlled by an overflow pipe, no further control of the separator is required. In some cases, however, float type level controllers may be used to control these two levels.

Oil from the separator is usually pumped into the surge tank and the water is pumped into the water storage tank. The separator pump and the water pump operate as long as there is oil flow out of the heater. These pumps are turned on and off by a flow switch, FSL 1.

The water content of the oil pumped from the surge tank is measured by an analyzer controller, AC 1. This controller positions a diverter valve, FCV 1, either to recirculate the oil containing excessive water or to allow the oil meeting the specification to go to the customer's storage tank. Because the water in the oil may also be mixed with very fine sand, called black sand, from the producing formation, the analyzer controller, AC 1, is called a basic sediment and water (BS and W) monitor. This device uses a capacitance probe to detect the amount of water in the oil and is accurately calibrated to the specified water content. The diverter valve is in the recirculate position whenever the main transfer pump is off. It is kept in that condition until a short time after the pump starts to allow the BS and W monitor to sense the actual water content of the oil in the tank.

A level controller, LC 1, on the surge tank can be used either to stop the main transfer pump or to recirculate the oil flow when the tank level is low. When the level returns to normal, the pump is restarted and the BS and W monitor takes over control of the diverter valve.

The volume flow of oil from the LACT unit is measured and digitally indicated by the positive displacement meter, FQI 1. Although the flow rate could be controlled by the tank level controller, LC 1, no flow rate control, except possibly a manually adjusted valve, is usually used. When it is in by the separator and tank, the oil is usually near ambient temperature. Therefore, temperature correction of the flowmeter, FQI 1, is not normally necessary.

Crude oil, as it leaves the LACT unit, still contains some black sand and water as well as paraffins and possibly some corrosive contaminants. Therefore, the BS and W monitor and the flowmeter must be constructed for accuracy and reliability under these condtions. All other controls are for the protection or operation of the equipment and must be reliable even though they may not be actuated frequently.

APPLICATION
DATA

When a remote monitoring system is used, the main transfer pump can be remotely controlled. Since the volume of oil from the LACT unit is the basis of payment to the operator, the indication of the flowmeter, FQI 1, is always stored locally as well as being read by a remote monitoring system. To indicate possible malfunctions, the high tank level, pump motors off, diverter valve position, pilot flame out, high heater water bath temperature and low water bath level are usually remotely monitored as alarm functions.

The well casing is connected to valves called *main valves*, through which the production tubing must pass. These valves are used to close in the well whenever the tubing is withdrawn for service. Valves connected to a tee above the main valves are used to inject or remove special mud which facilitates insertion or removal of the tubing. When these valves are not being used, they are tightly plugged by a bull plug. One of these bull plugs can also be equipped with a tap for measuring the casing pressure.

It is difficult to determine when a fault in the tubing, packers or casing occurs since it is not practical to place full-time monitoring devices in the well. Tubing and casing pressures are almost the only means of detecting such faults. On manually operated wells, these pressures are checked frequently and logged for comparison to previous values. When remote monitoring is used, these pressures are frequently monitored as deviations from preset levels or as actual values.

SEPARATORS

Condensate or crude oil wells usually produce a mixture of gas, hydrocarbon, and water. These three substances must be separated so that the water can be discarded and the oil and gas components be made available as separate product streams for further processing. This section considers some of the factors involved in effecting such separations, some typical separators, and their control schemes.

The relative amount of gas in the output stream from a well varies considerably. Gas lift wells produce large volumes of gas along with liquids. Natural lift wells also produce some gas which comes out

GAS CONTENT

of solution when the fluid pressure drops as it is brought to the surface.

The point in the production system where gas and liquid are separated depends on the relative ratios of these two phases. If large volumes of gas are produced, such as with gas lift wells, then the separation would probably be done at or near the well site. The gas and liquid are then piped separately to the central battery where the outputs from a number of wells are merged. For wells producing low or insignificant amounts of gas, such as natural wells with water drive, the well effluent can be piped directly to the battery where the separation of any small amounts of gas would take place.

WATER CONTENT

Water contained in the effluent from a well is often present largely as an emulsion dispersed in the hydrocarbon liquid phase. This emulsion must be broken down. Then gravity separation into water and liquid hydrocarbon phases is easily done.

These emulsions can be broken down either by heating the liquid or by the addition of emulsion-breaking chemicals. The heating method removes more water, but the water system is a complex and expensive device requiring more maintenance than a chemical system. The heating method is therefore not generally used unless necessary for some special condition.

The heaters generally provide indirect heating. An oil or gas

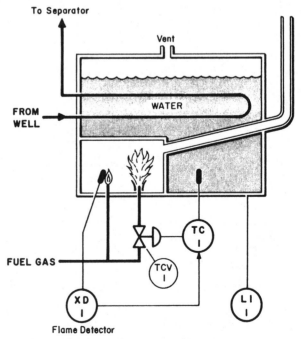

Figure 4—10. Typical oil heater.

flame heats the well effluent flowing through pipes immersed in the water bath. A safe method, this allows easier control than the use of a direct fired heater. A typical heater is illustrated in Figure 4–10.

A separator is basically a tank through which the well effluent flows. This constitutes an effective increase in the cross-sectional area of the pipeline, thus slowing fluid velocity and providing time for the various materials to separate under the influence of gravity. These tanks can be either vertical or horizontal cylinders. The former is generally used primarily to separate oil and water when smaller amounts of gas are present. Figures 4–11 and 4–12 illustrate the arrangements for vertical and horizontal separators respectively.

TYPES OF SEPARATORS

Controls are required to regulate variables associated with both the separator itself and, if used, with the heater needed to facilitate breaking of the water emulsion.

CONTROL SYSTEM

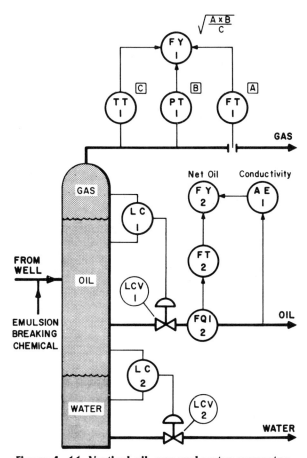

Figure 4–11. Vertical oil, gas and water separator.

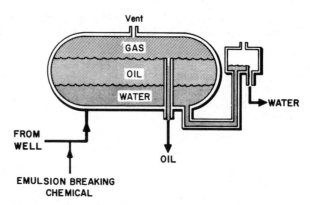

Figure 4–12. Horizontal oil, gas and water separator.

Separator Controls

Vertical and horizontal separators are controlled in the same manner. Variables to be controlled include interface levels, output volume metering, and flow rate for the addition of emulsion breaking chemicals. Figure 4–11 illustrates all three types of controls.

Interface levels are usually controlled by simple float actuated valves or by a gas actuated level controller sending a signal to a diaphragm actuated control valve. As shown in Figure 4–11, level controller, LC 1, regulates the gas-oil interface by manipulating a valve, LCV 1, thus controlling the rate of oil withdrawal from the midsection of the tank. Similary, level controller, LC 2, regulates the oil-water interface by acting on a valve, LCV 2, and controlling the rate of water withdrawal from the bottom section.

In some cases, a simple self-regulating scheme is used, as illustrated in Figure 4–12. Here, no mechanisms are needed since the oil-water interface is self-regulating by the use of a seal leg. The oil is regulated by a simple overflow pipe.

The gas and oil streams are metered at some point in the producing field, either at the well site or at a central battery. Gas flow is usually measured across an orifice plate in the flow line. Temperature and pressure may also be measured and used as correction factors to calculate the equivalent gas rate at standard temperature and pressure values. This compensation calculation can be mechanized to provide a single signal representing corrected flow rate to be recorded. It can also be integrated automatically over time to provide an indication of the accumulated total quantity of gas produced. This value can be displayed in digital form for convenient reading. It can be displayed locally at metering and it can also be transmitted for remote reading if a remote monitoring system is used.

Oil from the separator usually goes to a central battery for metering, but may sometimes be measured at the well site. Measurement is usually by means of a positive displacement type meter, FT 2.

A net oil computer, FY 2, measures the percentage of water in the oil by means of a capacitance probe. It also corrects the total volumetric flow to give the net oil flow. The output is displayed digitally and may also be transmitted to a remotely located indicator.

Other values which may also be indicated through a remote monitoring system are high or low levels in the separator or high water storage tank level. These signals would indicate the need for local service or other corrective action.

Emulsion-breaking chemicals are used in small quantities and control is usually not very critical. The simplest scheme is to add them periodically manually and to depend on the large separator volume to help smooth out variations to an acceptable level. Alternatively, the chemicals may be added continuously without automatic control, depending on a simple manual needle valve to set an approximate flow rate. In the case of large battery operations, a flow controller may be used, usually using a rotameter as the primary measuring element.

Heater Controls

Figure 4–10 shows a heater which may be used instead of chemicals to remove water from the oil. The temperature of the water in the heater is controlled by a temperature controller, TC 1, which uses an on/off function. The temperature of the oil is not critical, but does approach the temperature of the water. A pilot flame detector, FD 1, overrides the temperature controller to prevent opening of the control valve when the pilot flame is out. If this heater is remotely monitored, high and low water temperatures and pilot flameout are usually the displayed functions.

Application Data

It is beneficial to continue operation of the production system when the electrical power fails. Therefore, all control systems used for measuring product flow and determining payment must be functional at all times. Some type of mechanical record is usually made at the measurement site in this case.

The other controls on a separator or heater are usually either mechanical or gas actuated. If gas from the well is used, the condensates must be removed and the construction material of the controllers must be considered carefully since sour gas is corrosive.

INSTRUMENT LIST

	Tag No.	Description
Heater control system	XD 1	Pilot flame detector
	TC 1	Water temperature controller
	TCV 1	Fuel gas control valve
	LI 1	Water level indicator

	Tag No.	Description
Gas flow loop	TT 1	Temperature transmitter
	PT 1	Pressure transmitter
	FT 1	Flow transmitter
	FY 1	Gas flow computer—multiplier, integrator
Oil flow loop	FQI 2	Positive displacement flowmeter
	FT 2	Oil flow transmitter
	AE 1	Conductivity element
	FY 2	Net oil computer
Gas-oil interface level	LC 1	Level controller
	LCV 1	Oil flow control valve with electro-pneumatic transducer
Oil-water interface level	LC 2	Level controller
	LCV 2	Water flow control valve with electro-pneumatic transducer

NONASSOCIATED GAS PRODUCTION

Nonassociated gas normally occurs in formations which do not contain crude oil. The producing formation contains hydrocarbons and other impurities, all in a vapor state. This gas probably contains a very small percentage of hydrocarbons heavier than methane. This type of gas is produced primarily for its methane content which is commonly used as a fuel. Occasionally some heavier hydrocarbons can be economically recovered. One of the most common impurities in the produced gas is water since the gas was probably in direct contact with flowing water at some time in its formation.

The producing formation can be at pressures as high as 30,000 psi (210 000 kPa); the formation temperature is somewhat dependent upon its depth. Temperature gradients from the surface to the formation vary from 6° to 30°F (−14° to −1°C) per 1,000 feet (300 m) of depth. A 20,000 foot (6 000 m) well can be at 17,000 psi (120 000 kPa) and 400°F (200°C). The gas composition from this formation would be at less than liquid vapor equilibrium conditions for the hydrocarbons and water since it is not in contact with a liquid substance.

Gas produced from a group of wells in the same formation is collected and taken to a single pipeline or processing plant by a pipe network called a *gathering system*. Normally the operating pressure of the gathering system is about 1,000 psi (7 000 kPa) at about 50°F (10°C), which is ambient temperature for a buried pipeline.

Since the gathering system cannot accept formation pressures, pressure is reduced at the well head site. When gas expands, its temperature and pressure decrease. Under some operating conditions this causes the water and occasionally some of the heavier hydrocarbons to condense at the well head. It is undesirable to have condensed liquids in a gas pipeline. Therefore, a condensate separator is usually located at the well head site to remove the condensate before the gas enters the gathering system.

When the well is shut-in, the gas in the production tubing is at an equilibrium temperature with the surrounding formation and is much cooler than the bottom hole temperature. The well head pressure is at bottom hole pressure minus the hydrostatic head pressure of the gas in the pipe. This is the maximum pressure condition at the well head. When the well is opened, an increase in flow rate decreases the well head pressure due to pressure drops in the formation and the production tubing.

The high pressure drop and low initial temperature of the gas causes the produced gas to be much colder than if from a well which had been flowing at a high rate for some time. This lower temperature can freeze water condensed from the gas and cause plugging of critical control and protective equipment. To prevent freezing, heaters are usually installed at each well site to reheat the gas to normal temperature as soon after the pressure reduction point as possible.

Well head pressure is usually reduced through a choke valve. This valve is a special design, built to withstand the high well head pressure, high flow velocity and flow of condensate or ice forming inside the valve. The most severe conditions occur at this valve for a short time when the well is opened and during low production rates. At higher flow rates and when the well temperature has come into equilibrium, a more conventional valve will usually give satisfactory service since there will be less condensate and no ice formed. Since the heater is not required under these conditions, two valves are frequently used (see Figure 4–13). Flow is through the heater choke valve and heater during the most severe conditions and through the heater bypass valves as the conditions become less severe.

CONTROL SYSTEM

Gas Heater

The gas heater is usually a water bath type because of ease of operation and safety. The water is held at a nearly constant temperature by the on/off controller, TC 1. Gas passing through the tubes is heated by the water. The heater is sized to provide the heat required at all gas flow rate conditions. Since the main flame in the heater is periodically turned off, a pilot flame is provided for reignition. The temperature controller, TC 1, can be overridden by a pilot flame detector, XD 1, to prevent opening of the fuel valve when there is no pilot flame. The heater water level is always indicated on level indicator, LI 1, for periodic checking. This level

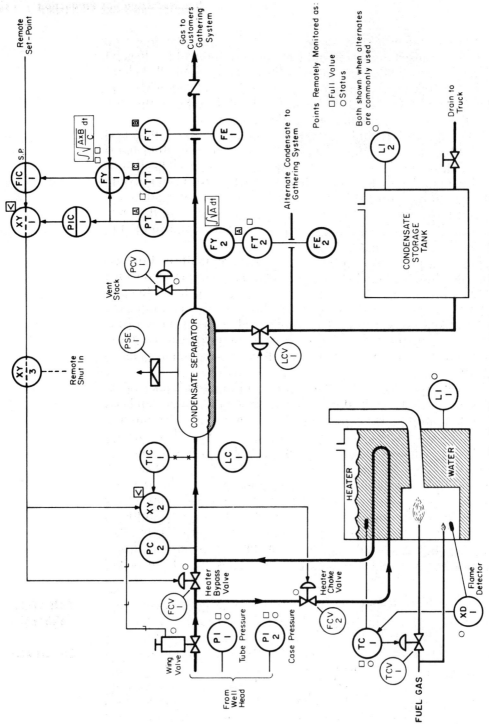

Figure 4–13. Condensate separator control system.

124

and other heater malfunctions may be interlocked with control functions to prevent operation of the well when one or more malfunctions exist.

If the well is to be operated by a remote monitoring system, an overall heater malfunction or individual heater malfunctions are remotely indicated. These individual malfunction indicators can be high or low water temperature, pilot flameout and low water level.

Condensate in the gas is removed by reducing the gas velocity in order to allow the entrained condensate to drop out. This is done in the condensate separator. A level controller, LC 1, is used to regulate the flow of condensate from the separator. **Condensate Separator**

Since the amount of condensate from a nonassociated gas well is usually small and primarily composed of water, the condensate from the separator is stored locally in a storage tank, as shown in Figure 4–13. A level indicator, LI 2, on this tank is used to show when a tank truck should be called to empty the tank. A record of the amount of condensate formed is kept as a measure of well performance. This is done by logging the amount of condensate removed by the tank truck. When the well is operated by a remote monitoring system, the high storage tank level is remotely indicated.

Occasionally the condensate is removed from the site by a condensate gathering system. In this case, the flow is metered and the volume is periodically checked. This value is also remotely indicated for periodic reading when a remote monitoring system is used.

Gas is sold on the basis of a flow rate agreeable to the producer and the customer. Maximum flow rates are determined by state or federal regulations or production company policy to obtain optimum recovery. The flow rate of the well is always monitored in terms of standard cubic feet (m^3). At the point where the gas passes into the purchaser's system, called *custody transfer*, the flow rate is measured as accurately as possible. The point of custody transfer can be at the well site if the purchaser owns the gathering system. An orifice plate, FE 1, and a D/P transmitter, FT 1, are the usual means of measuring the gas flow rate. The measured differential pressure is then corrected for the temperature and pressure variations in a mass flow computer, FY 1. The output of the flow computer is usually proportional to the flow rate in millions of standard cubic feet per day (millions of cubic meters per day), and the gas flow is usually integrated and totalized. When a remote monitoring system is used, both values are usually remotely indicated. Usually, the computed flow rate or the measured variables are recorded at the well site as proof of contract performance or of well performance. When **Gas-to-Gathering System** a remote monitoring system is used, these values are frequently locally recorded as record assurance in case of monitoring system failure.

Well Control

Normal control of the well is performed by a flow controller, FIC 1, with the output of the gas flow computer as the process signal. Since high pressure in the production equipment is possible, the flow control is usually provided with a pressure controller, PIC 1, override through a low signal selector, XY 1. This is primary protection against high pressure in the low pressure rated part of the system. A remote monitoring system can adjust the set point of the flow controller but is rarely used to adjust the set point of the pressure controller since this is basically a safety type of control system.

The heater choke valve and the heater bypass valve are positioned by the output of the low signal selector, XY 1, and through the signal selector, XY 3. Zero to fifty percent output from XY 1 strokes the heater choke valve from closed to fully open. Above fifty percent, the heater bypass valve starts to open and is fully open at one hundred percent. Either valve is capable of passing the required flow rate in its full open position. As the temperature rises to the set point of the temperature controller, TIC 1, the controller acts to close the heater choke valve through the low signal selector, XY 2. This removes heater choke valve control from the flow controller, FIC 1, and causes the output of FIC 1 to increase until the heater bypass valve opens. This makes up the flow loss through the heater choke valve. The temperature of the gas through the heater bypass valve can become warm enough to force the temperature controller to close the heater choke valve.

Well Shut-In

When the well is shut in through a remote monitoring system, the output to the valves is reduced to zero by means of a signal selector, XY 3, in response to the remote command. Since there can be some leakage through these valves in the closed position, the manual valves at the well head are closed for long-term shut-ins.

Protection from Overpressure System

Since very high pressures are possible on the production equipment, the primary pressure controller, PIC 1, is backed up with redundant pressure protection systems. A self-contained back pressure regulator, PCV 1, vents gas in the production equipment through a vent stack as the pressure rises above the control point of PIC 1. If this fails to reduce pressure, the wing valve is tripped closed by PC 2. This control requires manual resetting to open the wing valve. Finally, a rupture disk, PSE 1, prevents over pressure in the condensate separator. The condition of these pressure safety devices is usually indicated with a remote monitoring system.

Well head tube and casing pressures are important in the determination of the well condition. Local indication of these values is always provided. When a remote monitoring system is used, these values can be remotely indicated as a full value or deviation from a normal value.

Control equipment at the well site is frequently operated with gas as the energy source because electric power and compressed air are not readily available. This gas can be sour (i.e., contains hydrogen sulfide) so care must be taken in the material selection used in contact with this gas, even though the gas can be treated to remove most of the corrosive material.

If a remote monitoring system is used, reliable electric power must be available at each well site to operate the control equipment. Gas will be used to operate the valves, the electro-pneumatic transducer and the valve actuator. These must be selected for corrosion resistance.

When the field terminal of a remote monitoring system is centrally located for a group of wells, it is most practical to have the monitored points at each well brought to this central location by wire rather than tubes because of the distances involved. Here the most common practice is to locate the flow controller and computer at the central location. This equipment could then be electronic. All other control equipment would be at the well site and gas operated, except the three electronic transmitters, PT 1, TT 1 and FT 1. All other signals would be communicated to the central location by wire as status conditions. The pressure controller, PIC 1, is a safety device and is usually located at the well site although this controller and the signal selectors, XY 1 and XY 3, could be centrally located without loss of protection. The temperature controller, TIC 1, is commonly located at the well site since it does not have direct connection to the monitoring system and this arrangement requires two less signal connections.

In most override control systems, external feedback from the signal selector output to the associated controllers is advantageous in preventing overpeaking of controllers as they override each other. Therefore, the feedback signal from the signal selector, XY 2, would be connected to the temperature controller, TIC 1.

When the pressure and flow controllers, PIC 1 and FIC 1, are located together, the feedback signal from the signal selector, XY 1, is connected to both controllers. If the pressure and flow controllers are not in the same location, the signal selector will probably be in the same location as the pressure controller. The feedback signal from the selector should be connected to this controller. The absence of external feedback to the flow controller is not serious in this system.

Load changes in most gas well systems are not rapid and recovery from the load or set point changes should be without overpeaking or oscillation. Slow recovery of the control system, except for protective devices, is preferred. Therefore, the response of the flow controller in particular is overdamped.

INSTRUMENT LIST

	Tag No.	Description
Heater control system	XD 1	Pilot flame detector
	TC 1	Water temperature controller
	TCV 1	Fuel gas control valve
	LI 1	Water level indicator
Well head control components	PC 2	Feed gas pressure controller
		Wing valve-feed gas pressure control valve
	PI 1	Well tube pressure indicator
	PI 2	Well case pressure indicator
Condensate storage control components	LI 2	Condensate level indicator
Condensate separator control components	LC 1	Condensate level controller
	LCV 1	Condensate level control valve
	PSE 1	Separator pressure relief
	FY 2	Condensate meter
Gas flow control components	FT 1	Gas flow transmitter
	TT 1	Gas temperature transmitter
	PT 1	Line pressure transmitter
	FY 1	Gas flow computer
		Multiplying computer, dividing computer, square root extracting computer, integrator
	FIC 1	Gas flow controller
		Set-point transducer
	PIC 1	Line pressure controller
	XY 1	Low signal selector
	HY 3	Remote shut-in relay with electro-pneumatic transducer
	TIC 1	Gas temperature controller
	XY 2	Low signal selector
	FCV 1	Heater bypass valve
	FCV 2	Heater choke valve
	PCV 1	Gas pressure regulator

GAS TREATMENT

The gas from natural gas fields usually contains materials or contaminants other than the required hydrocarbons. Some contaminants (water and sulfur compounds like hydrogen sulfide, mercaptans and traces of other organic sulfides) create difficulties in handling the produced gas.

Water is present in nearly all of the produced gas. At pipeline pressures, water forms hydrates which accumulate like packed snow and cause a partial or complete blocking of the pipeline.

Hydrogen sulfide (H_2S) causes corrosion and has a very strong odor. Other contaminants, such as carbon dioxide (CO_2) and nitrogen (N), are undesirable because they decrease the heating value of the gas. Helium (He), recovered for its value, may be present in the natural gas up to about 0.3 percent.

WATER REMOVAL PROCESS

Gas from a producing field can contain from as little as three pounds to as much as 1,400 pounds (1.0 to 700 kg) of water per million cubic feet (MMCF) (28 300 m^3) of gas. To prevent hydrate formation in the pipeline, the water content of the gas must be reduced to below the dew point of water at the highest operating pressure and lowest operating temperature of the pipeline. The equilibrium moisture content of natural gas at 1,000 psi (6 900 kPa) and 50°F (10°C) is 12.9 pounds (5.85 kg) of water per MMCF (28 300 m^3). The usual requirement for dried gas is 6 pounds (2.7 kg) of water per MMCF (28 300 m^3).

Water can be removed from the gas by compression and cooling, liquid absorption, solid absorption and solid desiccants. The drying process can involve more than one of these methods. Liquid absorption by diethylene glycol (DEG) or triethylene glycol (TEG) is the most commonly used process. Although DEG is still used in many older systems, TEG is presently favored because of its higher thermal decomposition temperature and lower vaporization loss.

A typical TEG liquid absorption system is shown in Figure 4–14. Wet gas at approximately 500 psi (3 450 kPa) passes upward through a gas-liquid contactor which is similar in construction to a bubble cap distillation column. TEG, containing very little water, is fed into the top of the column and passed downward. The dew point of the dried gas is a function of the water in the feed TEG, the circulation rate per MMCF (28 300 m^3) of gas, the water content of the feed gas and the contactor temperature.

The TEG from the bottom of the contactor contains water absorbed from the gas. This water must be removed before the TEG can be recirculated. This is done by a distillation column called a *regenerator*. With the regenerator at atmospheric pressure and the reboiler temperature at 400°F (204°C), a bottom product TEG containing about one percent water is obtained.

The overhead product of the regenerator is water vapor. Part of this is condensed and flows back into the column as reflux and the remainder of the vapor is vented. An air fin radiator is often used as an overhead condenser when the regenerator is operated at atmospheric pressure.

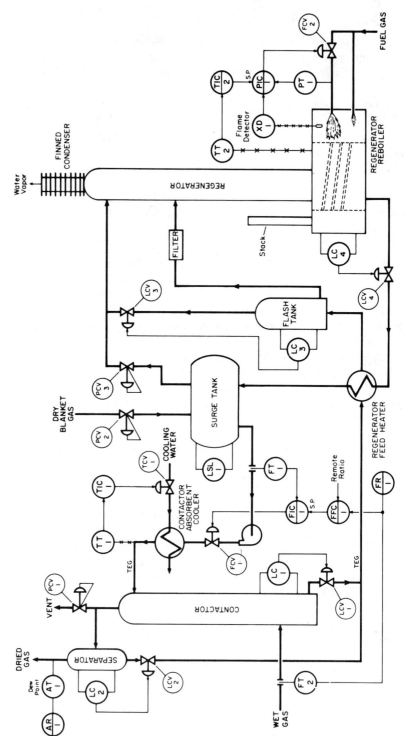

Figure 4–14. Typical triethylene glycol gas drying control system.

130

The dry TEG leaving the regenerator must be cooled to the optimum contactor operating temperature. Part of the cooling is done by using the regenerator bottoms to heat the regenerator feed stream. Some heat is lost by the regenerated TEG while it is in the surge tank. The remaining heat is removed in a water cooled heat exchanger.

Various steps are taken to maintain TEG inventory in the system. A separator recovers the TEG that is entrained in the gas from the contactor. Some of the absorbant is lost by evaporation in both the contactor and the regenerator. A decrease in the surge tank level is the usual indication of this loss and fresh TEG can be added as required. To prevent absorption of atmospheric moisture and oxygen (which would be released into the gas stream), the surge tank is usually blanketed with gas at low pressure.

The wet feed gas to the contactor can contain condensed liquids such as water or hydrocarbons. These liquids are usually removed by a separator before they reach the contactor. This removal reduces the load on the regenerator and recovers hydrocarbons which would either pass out of the regenerator as vapor or remain in the absorbant as contaminants.

The regenerator reboiler can be heated either by direct firing, as shown in Figure 4–14, or by steam, depending upon availability. As most gas driers are not located at a major plant site, steam generation is not practical. When direct firing is used, the regenerator is located a safe distance from the contactor.

CONTROL SYSTEM

The water content of the treated gas is a function of the water content of the feed gas, the circulation rate of the absorbant per MMCF (28 300 m^3) of gas, the quality of the absorbant and the contactor temperature. The quality of the absorbant to the contactor is controlled by controlling the regenerator reboiler temperature at the established pressure. Since the back pressure to a direct fired burner (see Figure 4–14) is a measure of the fuel feed rate, a cascade control system consisting of a reboiler temperature controller, TIC 2, and a fuel pressure controller, PC 1, is used to regulate the reboiler temperature.

Regenerator overhead vapor condensation is not usually controlled. In the TEG system, the amount of condensation for reflux is not critical and variations due to ambient changes do not significantly change the regenerator operation. Usually the overhead condenser water flow rate is set to handle maximum load conditions. If cooling water flow is economically important, the condenser water flow rate could be controlled to produce condensate several degrees below the normal equilibrium temperature.

Temperature control of the absorbant to the contactor is obtained by adjusting the cooling water rate to the contactor absorbant cooler. Since this temperature is important, a temperature controller, TIC 1, is used to measure the absorbant temperature and adjust the cooling water flow.

The ratio of absorbant flow to gas flow in the contactor is a function of the amount of water in the feed gas. Usually treating units are used to handle gas from a single field. The flow rate can vary but the water content will stay nearly constant. Therefore, the ratio of absorbant flow, measured by the flow transmitter, FT 1, to the gas flow, measured by the flow transmitter, FT 2, can be set on the ratio station, FFC 1. The absorbant flow rate controller, FIC 1, controls the absorbant flow at the required ratio to the gas flow.

A dew point recorder, AR 1, is often used to establish the flow ratio necessary for the required gas composition. In rare cases where the inlet gas composition varies, this measurement could be used to adjust automatically the flow ratio. These recorders are also used to indicate decomposition or contamination of the absorbants or other malfunctions in the system.

The gas drying and gas treating processes can run for long periods of time with only periodic inspection. If the gas field is remotely monitored, some variables which indicate the process performance or a possible malfunction can be remotely monitored. The only variable which can be remotely adjusted is the absorbant to gas flow rate.

Application Data

Electric power is usually available for the operation of electronic instruments. Gas instead of air is commonly used to operate pneumatic instruments. The gas is sweet and no special consideration with respect to corrosion is necessary.

Some of the process variables can be recorded locally, particularly when remote monitoring is not used. In some cases, a small weatherproof building is used to house some of the recorders or controllers.

INSTRUMENT LIST

	Tag No.	Description
Contactor control system	FT 1	TEG flow transmitter
	FIC 1	TEG flow controller
	FR 1	Gas flow recorder
	FT 2	Gas flow transmitter
	FFC 1	Flow ratio relay with electro-pneumatic transducer
Level control components—bottoms	LC 1	Level controller
Level control components—tops	LC 2	Level controller

	Tag No.	Description
Analyzer control components	AT 1	Analyzer transmitter
	AR 1	Analyzer recorder
Cooler control components	TT 1	Temperature transmitter
	TIC 1	TEG temperature controller
Surge tank control components	LSL 1	Level switch low
Regenerator control system		
Fuel gas control components	TT 2	Temperature transmitter
	TIC 2	Temperature controller
	XD 1	Flame detector
	PT 1	Pressure transmitter
	PIC 1	Pressure controller
	LC 4	Level controller with electro-pneumatic transducer
Flash tank control system	LC 3	Level controller

H_2S and CO_2 are removed from the gas by chemical reaction with an aqueous amine solution. The process and instrument diagram is illustrated in Figure 4–15. A 15 to 20 percent by weight monoethanolamine (MEA) or 20 to 30 percent diethanolamine (DEA) solution in water is used to react with the H_2S and CO_2.

H₂S AND CO₂ REMOVAL PROCESS

Sour gas from the well can contain up to about 45 percent by volume H_2S and usually from 0.1 to 6 percent CO_2. Some wells produce as much as 100 percent CO_2.

An acceptable level of H_2S for distributed gas is about one grain (7,000 grains = one pound, 1 grain = 0.0648 gram) per hundred cubic feet (CCF) (2.83 m^3) of gas. CO_2 content of the gas is not critical and can be whatever is needed to achieve the required heating value. Gas is not normally treated for the removal of CO_2 alone.

H_2S reacts with the amine solution at temperatures below 100°F (38°C). The CO_2 reacts best at a temperature of about 120°F (49°C). The reverse reaction, which strips the H_2S and CO_2 out of the amine solution, eliminates almost all of the H_2S at 240°F (116°C), but leaves 0.5 to 1 cubic foot (0.014 to 0.028 m^3) of CO_2 per gallon (3.79 l) of solution.

Water loss from the MEA solution is detrimental. Gas enters and leaves the contactor at saturated conditions so water loss from the regenerator overhead is prevented by using a water cooled condenser. This returns all the water to the regenerator as reflux. A slight positive pressure is held on this regenerator to provide a force to drive the H_2S and CO_2 gases to the flare or sulfur recovery unit. The regenerator reboiler can be direct fired or steam heated. A steam heated system is shown in Figure 4–15.

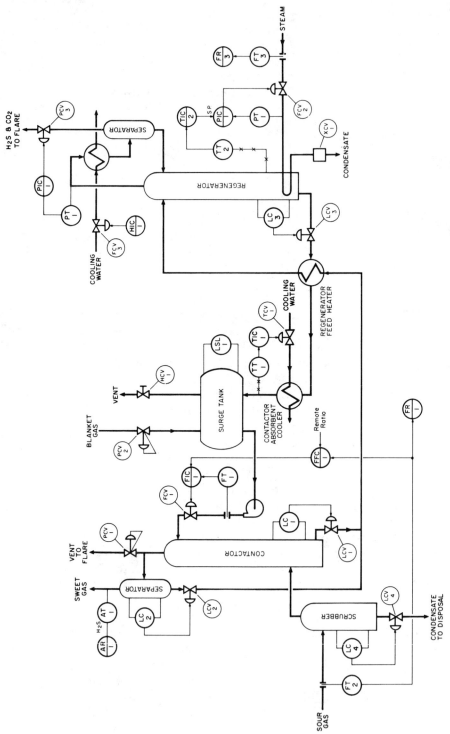

Figure 4–15. Typical monoethanolomine gas treating control system for removal of H$_2$S and CO$_2$.

Nitrogen and helium are not usually removed from the gas unless the helium is to be recovered. The separation is effected by distillation at low temperature and moderate pressure. This is usually done as part of the operation in gasoline plants which use the same type of process for separation of hydrocarbons.

NITROGEN AND HELIUM REMOVAL

The H_2S content of the treated gas is a function of the H_2S content of the feed gas, the circulation rate of the absorbant per MMCF of gas, the quality of the absorbant and the contactor temperature.

CONTROL SYSTEM

The quality of the absorbant to the contactor is controlled by regulating the regenerator reboiler temperature at the established pressure. A cascade control system consisting of a reboiler temperature controller, TIC 2, and a steam pressure controller, PC 1, is used to control the reboiler temperature (see Figure 4–15).

Water loss in the MEA system is critical, and total water condensation is required. Usually the overhead condenser water flow rate is set to handle maximum load conditions. If cooling water flow is economically important, the condenser water flow rate can be controlled to produce condensate several degrees below the normal equilibrium temperature.

Temperature control of the absorbant to the contactor is obtained by adjusting the cooling water flow rate to the contactor absorbant cooler. Since this temperature is important, a temperature controller, TIC 1, is used to measure the absorbant temperature and to adjust the cooling water flow.

The ratio of absorbant flow to gas flow in the contactor is a function of the amount of H_2S in the feed gas. Usually treating units treat gas from a single field. The flow rate can vary but the H_2S content will stay nearly constant. Therefore, the ratio of absorbant flow, measured by the flow transmitter FT 1, to gas flow, measured by the flow transmitter FT 2, can be set on the ratio station, FFC 1. The absorbant flow rate controller, FIC 1, then controls the required ratio of absorbant flow to gas flow.

A H_2S analyzer recorder, AR 1, establishes the flow ratio necessary to obtain the required gas composition. In rare cases where the inlet gas composition varies, this measurement could be used to adjust automatically the flow ratio. These recorders also indicate decomposition or contamination of the absorbants or other malfunctions in the system.

The gas drying and gas treating processes can run for long periods of time with only periodic inspection. If the gas field is remotely monitored, some variables which indicate the process performance or malfunction may be remotely monitored. The only variable which can be remotely adjusted is the absorbant to gas flow ratio.

Application Data

Electric power is usually available for the operation of electronic instruments. Gas instead of air usually operates pneumatic instruments. The gas is sweet and no special consideration with respect to corrosion is necessary.

Some of the process variables can be recorded locally, particularly when remote monitoring is not used. In some cases, a small weatherproof building is used to house some of the recorders or controllers.

INSTRUMENT LIST

	Tag No.	Description
Contactor control system		
Flow control components	FT 1	Absorbant flow transmitter
	FIC 1	Absorbant flow controller
	FT 2	Sour gas flow transmitter
	FFC 1	Multiplier
	FR 1	Sour gas flow recorder with electro-pneumatic transducer
Level control components	LC 1	Level controller
Separator control components	LC 2	Level controller
Scrubber control components	LC 4	Level controller
Analyzer control components	AT 1	H_2S analyzer transmitter
	AR 1	H_2S recorder
Surge tank control components	LSL 1	Level switch low
Contactor absorbant cooler	TT 1	Temperature transmitter
	TIC 1	Temperature controller
Regenerator control system		
Steam control components	TT 2	Temperature transmitter
	TIC 2	Temperature controller with electro-pneumatic transducer
	PIC 1	Pressure controller
	PT 1	Pressure transmitter
Steam flow components	FT 3	Flow transmitter
	FR 3	Flow recorder
	XCV 1	Steam trap
Level control components	LC 3	Level controller
Pressure control components	PT 2	Pressure transmitter
	PIC 2	Pressure controller with electro-pneumatic transducer
	HIC 1	Manual loader

ABSORPTION OIL PROCESS

Natural gas is a mixture of methane, ethane, propane, and other paraffinic hydrocarbons, along with H_2S, CO_2, N_2, He and traces of other compounds and elements. Normally, natural gas is processed to recover components heavier than methane, which is most often used for fuel. A common technique for releasing methane is that of lean oil absorption.

COMMON
TERMS

- **Flash vaporization.** An operation in which a liquid is heated and exposed to a lower pressure to convert a substantial portion of it to a vapor. Flashing is often done in a vessel called a *flash drum.*

- **Lean absorption oil.** An oil used to separate the heavier components from a vapor mixture by absorption of the heavier components during intimate contacting of the oil and vapor.

- **Mole (or mol)%.** The mole fraction of a component is the ratio of the number of molecules (or moles) of that component to the total number of molecules present in a mixture. By multiplying the mole fraction by 100, the mole % is obtained. Mole % may be acquired directly from process chromatograph analyses.

- **Presaturation.** Normally, the process of selectively attaching to a large molecular structure a simpler structure, making available positions for attachment of only larger structures at an ensuing exposure. Lean oil is presaturated with methane to enhance C_2 + absorption efficiency.

- **Residue gas.** Primarily methane (abbreviated as C_1). It is the gas product off of the absorption tower (absorber).

- **Rich oil.** Absorption oil containing dissolved natural gasoline fractions. Natural gasoline as used here describes a liquid containing primarily ethane (abbreviated as C_2) through the pentanes derived from natural gas.

- **Sponge oil.** An oil heavier than absorption oil used to minimize absorption oil carryover by gas entrainment.

- **Still (stripper).** A simple distillation tower used to remove vapors from liquids. The rich oil still separates pentanes and lighter hydrocarbon vapors from absorption oil.

- **Vol (volumetric)%.** The ratio of a pure component volume to the total mixture volume multiplied by 100. Chromatograph measurements are often given in vol %.

- **Wt (weight)%.** The ratio of a component's weight to the total mixture weight, multiplied by 100.

LEAN OIL ABSORPTION

The absorption oil process provides a low ethane recovery technique for separating ethane from natural gas. Recent improvements in refrigeration technology make it more feasible to chill lean absorption oil to permit greater ethane recovery. The basic lean oil absorption process is shown in Figure 4–16. Chilled natural gas is introduced into the bottom of the absorber, where the gas is contacted counter-currently by a presaturated lean absorption oil (lean oil). The lean oil absorbs ethane and heavier hydrocarbons (abbreviated as C_2^+), thereby producing a rich oil that accumulates at the tower bottom. The methane escaping overhead is directed to a gas transmission pipeline. When good ethane recovery is desired and, consequently, lean oil is refrigerated, the feed gas must be dehydrated to prevent plant freeze-ups.

The rich oil from the absorber is directed into the rich oil demethanizing (R.O.D.) tower where absorbed methane is removed by fractionation and exits from the top of the tower. The methane is further refrigerated, combined with regenerated lean oil, and sent to the presaturator. The vapor phase containing mostly methane may then be combined with the gas from the absorber overhead, or it may be used as plant fuel. The presaturated liquid is used for reflux in the R.O.D. tower and as lean oil to the absorber. For increased tower-top separation efficiency, a chilled pumpback stream may be added near the top of the R.O.D. tower.

R.O.D. bottoms are sent to the rich oil still, where the desired hydrocarbon components are separated from the absorption oil. The overhead vapors containing ethane and heavier components are partially condensed and collected in the reflux accumulator. The vapor from the accumulator is combined with the liquid not necessary for still refluxing and passed to the make tank. Here, the C_2^+ product is divided into recycle and charge streams to a light ends recovery unit. The still bottoms liquid is sent through a series of heat exchangers on its return to the presaturator.

Reboiler heat is supplied by either hot oil or steam. Tower reflux is generated by using refrigeration, cold water, or air coolers. Gas turbines are often used to drive compressors.

CONTROL SYSTEM

A. *Lean Oil Split.* By stoichiometry, the quantity of lean oil necessary for absorption of the C_2^+ hydrocarbons in both the absorber and the R.O.D. can be determined. Consequently, the flow of lean oil to the absorber should be set at a ratio of the inlet gas flow to a given gas composition (see Figure 4–16). This same principle and control pertain to the flow of sponge oil to the absorber and lean oil reflux to the R.O.D. However, since fluctuations of lean oil composition and R.O.D. operation occur, a compensation of the reflux flow should be made accordingly.

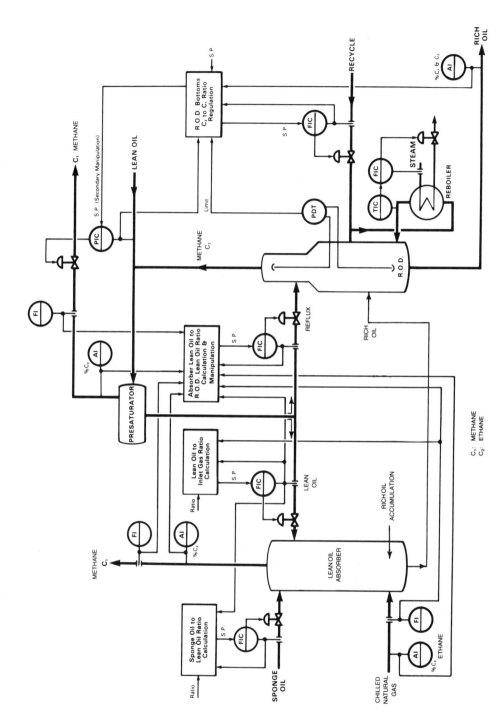

Figure 4–16. Control scheme for maximum ethane recovery.

This is accomplished by making small manipulations in the ratio of the absorber lean oil flow to the R.O.D. lean oil flow in a manner to maximize ethane recovery. Ethane recovery is determined by volume balance.

B. *R.O.D. Bottoms Composition.* Another indicator of ethane recovery is the mole ratio of ethane to methane in the R.O.D. bottoms. Shown in Figure 4–16, the C_2 to C_1 ratio should direct the adjustment of the C_2^+ recycle flow rate. Should the recycle flow rate reach a high constraint, the R.O.D. pressure should be lowered slowly to bring the effectiveness of the recycle flow back into perspective. The working range of the pressure is limited by constraints, such as tower boilup, maximum heat input, and compressor efficiency.

C. *R.O. Still Feed Preheat.* To stabilize the still's feed heat input, a continuously updated heat transfer rate calculation should be used to adjust the flow of the preheater's heating medium (usually steam).

D. *R.O. Still Reflux.* Since the still is used to split hydrocarbons lighter than the hexanes from absorption oil, the system is obviously multicomponent. Therefore, temperature cannot be used as an indicator of quality. However, reflux rate is nearly proportional to the charge rate given a constant still bottoms temperature profile. This condition permits the reflux rate to be adjusted at a ratio to the charge rate as shown in Figure 4–17.

Disturbance variables, such as foul weather and daily temperature changes, hinder the stability of the top of the still. These ambient temperature effects should be opposed by external reflux compensation.

E. *R.O. Still Bottoms Heat.* Unlike a classical distillation where a large portion of the tower feed is vaporized upon approach and rises through the tower, most of the still's feed falls to the bottom, where it absorbs a large quantity of heat. Again referring to Figure 4–17, a configuration of the reboilers and preheater such as that shown implies that a rather distinct profile of the bottoms temperature may be attained. For a given still feed composition, a fixed bottoms temperature profile will yield a specific slippage of pentanes to lean oil. Therefore, the heat inputs of the heat exchangers may be controlled as ratios to each other to define a temperature profile. Ratio adjustments will redefine the temperature profile, thus providing a change in the composition of the lean oil. The temperature profile should be adjusted in a manner to contain the slippage of pentanes to X%.

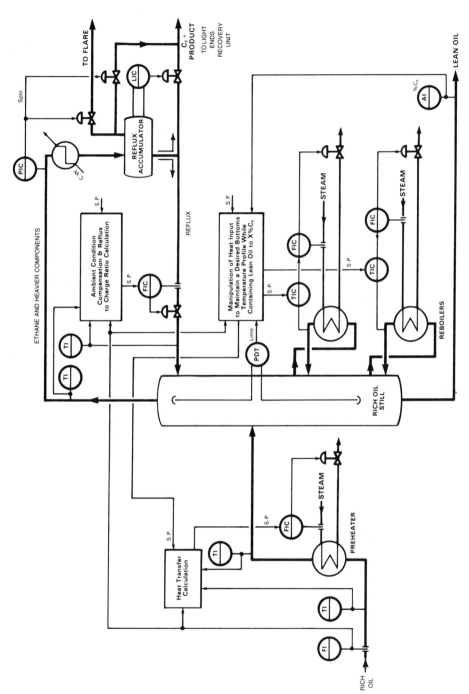

Figure 4–17. Rich oil still control scheme.

141

CRUDE OIL DISTILLATION

Crude oil is made up of an almost infinite number of discrete hydrocarbons, from methane (1 carbon atom) to materials having 70 or more carbon atoms. The first step in any petroleum refinery is the separation of the crude oil into various fractions by the process of distillation. These fractions may be marketable as is, or may be feedstocks for other refinery or processing units. In most refineries, the process is carried out by two distillation units: one operating at nearly atmospheric pressure and the other under vacuum.

COMMON TERMS

- **ASTM distillation** A laboratory distillation of petroleum products and intermediate fractions (cuts) to characterize their compositions in terms of boiling point range. The procedures for this controlled analysis were developed by the American Society for Testing Materials (ASTM).

- **Crude assay** The complete and definitive analysis of a crude oil.

- **Cut point** The boiling point division between cuts of a crude oil or base stock.

- **End point (EP)** The highest temperature of a boiling point range.

- **Flash zone** The area in a vessel where the lighter fractions vaporize or flash and the heavy oils drop to the bottom.

- **Initial boiling point (IBP)** The boiling point of the lightest hydrocarbon within a boiling point range.

- **Overflash** An extra amount of vaporization to insure that adequate reflux will be available in the trays between the flash zone and the lowest side-cut product draw tray. It is measured as a volume percent of crude oil charged to the tower, typically between 2 to 5 volume percent.

- **Pumparound** A system on a distillation tower for withdrawing a liquid from a tray, cooling it, and returning it to a higher tray for the purpose of inducing vapor condensation.

- **True boiling point (TBP) distillation** A complex laboratory distillation of crude oil to determine the relationship of the actual boiling point to the volume percent vaporized. The TBP range is indicative of actual composition. However, designers and plant operations personnel most often use ASTM boiling point range information.

ATMOSPHERIC DISTILLATION

Raw crude oil is first distilled at atmospheric pressure in the crude unit. The products from the crude unit range from a light straight-run fraction of up to C_6 hydrocarbons to a heavy bottoms residue requiring further distillation.

Prior to its introduction to the crude unit, the raw crude is washed in a desalter to reduce the amount of inorganic salt present. This minimizes corrosion of the refining equipment.

The heart of the crude unit is the atmospheric (crude) tower. The crude is heated to a temperature at which it is partially vaporized and then introduced to the tower at a point near the bottom (see Figure 4–18). The tower is equipped with a series of trays through which the vapors can pass in an upward direction. The vapors pass through a liquid level contained on each tray until they come to equilibrium with the liquid on a tray. However, they may continue all the way up and go overhead. The liquid flows downward continuously by gravity from tray to tray. As the vapors pass upward through the succession of trays, they become lighter (lower in molecular weight and more volatile), and the liquid flowing downward becomes progressively heavier (higher in molecular weight and less volatile). This countercurrent action results in fractional distillation, or separation of hydrocarbons based on their boiling points. A liquid can be withdrawn from any preselected tray as a net product. The lighter liquids are from trays near the top of the tower and the heavier liquids from the trays close to the bottom. Strippers located near the crude tower are used to increase separation efficiency.

A common product boiling point range profile is illustrated in Figure 4–19. The example at the bottom characterizes a crude tower's operating parameters.

A. *Desalter Level.* Desalters are very long, horizontal vessels containing a large volume of crude oil and water. The oil/water interface level in a desalter is typically measured by a buoyancy-type level transmitter, which has been calibrated for normal operating conditions.

Control System

Since a small change in level makes a very large change in liquid volume, it is essential that the interface level be controlled tightly to maintain a constant desalter throughput. Changes in the crude oil and water densities cause error in the output of the level transmitter, thus causing the actual interface level to swing uncontrollably.

To alleviate this error, the measurement must be density-compensated by an inferred density relationship temperature. This compensation aids in stabilizing the unit's water inventory by providing an indication of the true interface level for control. Consequently, oil carry-over to water treating and water vapor carry-over to the crude tower are minimized or eliminated.

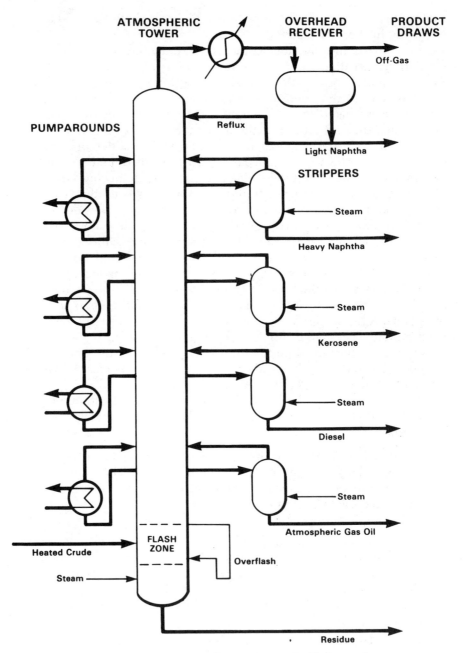

ATMOSPHERIC TOWER

OVERHEAD RECEIVER

PRODUCT DRAWS

Off-Gas

PUMPAROUNDS

Reflux

Light Naphtha

STRIPPERS

Steam

Heavy Naphtha

Steam

Kerosene

Steam

Diesel

Steam

Atmospheric Gas Oil

FLASH ZONE

Heated Crude

Overflash

Steam

Residue

Figure 4–18. A typical crude distillation unit.

144

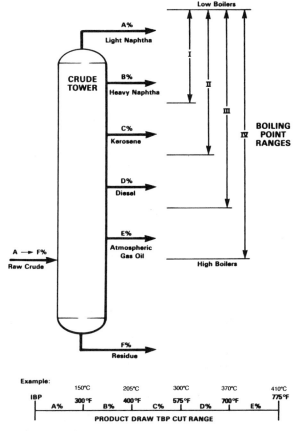

Figure 4–19. Product boiling point range profile.

B. *Heater Pass Flow.* The high output crude heater generally has a minimum of four passes, each with flow control to provide a means of thermal load balancing. Use of symmetrical piping by itself to distribute the flows equally is not sufficient, especially if some fouling has occurred.

   Conventionally, the operator makes occasional set point adjustments to the pass flow controllers in an attempt to produce equal outlet temperatures. These manipulations must be made slowly, consistently, and accurately, since altering one flow will inevitably upset the others (see Figure 4–20).

C. *Heater Combustion.* Since the crude heater consumes a great quantity of fuel, maximizing its efficiency is most desirable. Heater equipment such as draft transmitters, motor-operated dampers, and heater temperature, profile thermocouples are necessary for automatic trim control.

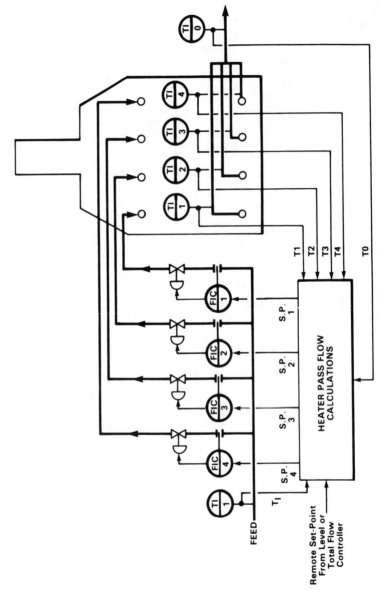

Figure 4–20. Heater thermal load balancing control.

The primary step in increasing heater performance is to maintain the design air-to-fuel ratio, allowing just a minimum of residual oxygen to insure complete combustion. Too much air increases the energy carried away in hot flue gases. Oxygen and carbon monoxide monitors on the stack provide combustion quality (stoichiometric) information.

Occasionally, inefficient firing control can be attributed to rapid, unpredictable changes in the fuel's composition, thus causing changes in the fuel's heating value. In this instance, it would be wise to add fuel heating value-shift compensation.

D. *Tower Pumparound.* The tower's temperature profile is primarily controlled by managing the heat removal above each product draw tray. The heat transfer rate is regulated by manipulating the pumparound flow rate.

To stabilize product separation, heat transfer rates should be constrained to a set value, given a steady state condition. Pumparound flow rates determined by heat transfer calculations will provide each control (see Figure 4–21).

The pumparound's heat removal allocations can be established by reviewing economic considerations and both physical stoichiometric constraints. Tower studies may indicate that tray loading calculations could set the heat transfer rates for optimal heat removal. However, it is usually more practical to use a scheme that will maximize heat change to the crude up to the physical limits.

In addition to (or instead of) providing the atmospheric gas oil (AGO) pumparound with heat transfer control, it may be advantageous to manipulate the pumparound flow between stream color constraints to minimize overflash. Color analyzers are necessary to detect high and low color limits.

E. *Tower Side-Cut Product Draw.* By the law of volume balance, there is interaction of product flow controllers in a multiproduct distillation column, as is the case with a crude tower. The product flow controllers must be decoupled to minimize product loss during a tower upset.

To achieve this decoupling, each side-cut product flow controller must be given a set point representing the ratio-to-crude of the total material product flow below that draw's specified end point. By this manner, a change in flow rate of a lighter product stream will be rapidly compensated by the next heavier product's total flow controller, without allowing the disturbances to affect heavier products (see Figure 4–22).

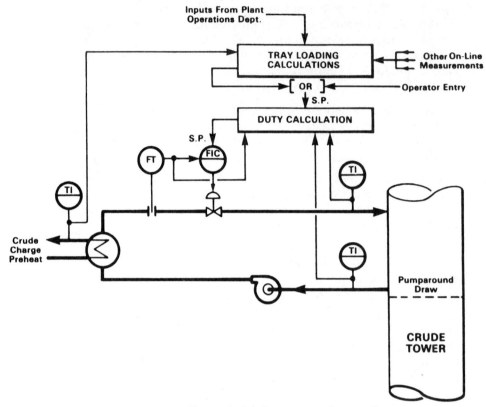

Figure 4–21. Pumparound control.

End point controllers should provide draw set points. End point analyzers or laboratory results may be used to input end point measurements. Viscosity (pour point) constraints may be added.

Referring again to Figure 4–19, it is obvious that by characterizing crude charge in terms of volume percent per boiling point range, product draw rates can be determined solely by volume balance.

  F. *Tower Overhead Vapor Temperature.* Stabilizing the top of the tower minimizes both product draw and deviations in overhead vapor composition. Two compensations are necessary to accomplish this stabilization. They entail pressure correction of the temperature measurement and automatic reflux rate adjustment in order to oppose ambient climate effects (see Figure 4–23).

A feedback composition control to adjust overhead vapor temperature for maintaining a consistent light naphtha quality may be beneficial, depending upon the marketing needs of the re-

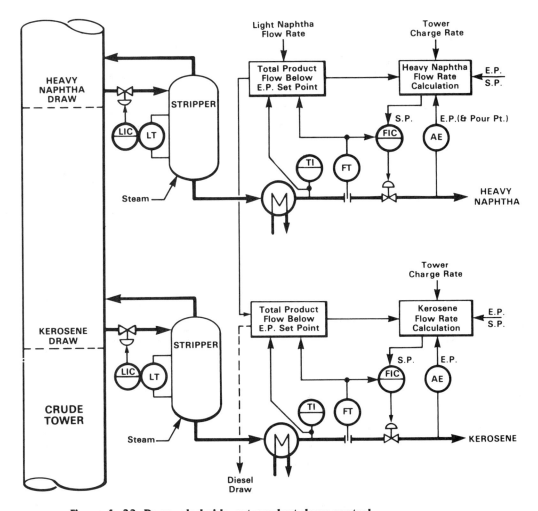

Figure 4–22. Decoupled side-cut product draw control.

finer. The vapor's composition may be monitored by a chromatograph or by a boiling point range analyzer.

G. *Feedforward.* Tower upsets caused by crude charge rate changes can be reduced by implementing feedforward adjustment of draws and pumparounds.

VACUUM DISTILLATION

The residue from the atmospheric tower (also termed *reduced crude*) may be further distilled by vacuum flashing. The residue is heated and introduced into a vacuum operated at near vacuum, maintained by steam ejectors or jets (see Figure 4–24). In this tower, a flash separation produces (vacuum) gas oils and nondistillable asphalt or pitch, the selection of which depends on operational re-

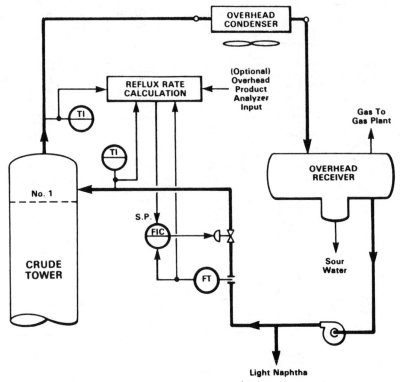

Figure 4–23. Overhead vapor temperature control.

quirements. More complex vacuum towers are used for the manufacture of lube oils or other types of specialty oils that necessitate a number of side-cut distillate draws.

Control System

A. *Heater Pass Flow.* It is the same as that for the crude oil distillation system.

B. *Heater Combustion.* It is the same as that for the crude oil distillation system.

C. *Tower Pumparound and Pumpback.* Because there are numerous vacuum tower designs resulting from different operational needs, it is impractical to propose a standardized control scheme. However, as a minimum plan, the reflux rates should be controlled to contain heat transfer rates at set values to augment tower stability. The set points may be simply operator entries or they may be directed by analyzer and calculated variable inputs.

D. *Feedforward.* Feedforward manipulation of reflux rates to compensate for charge rate changes can reduce any disturbance in tower operation.

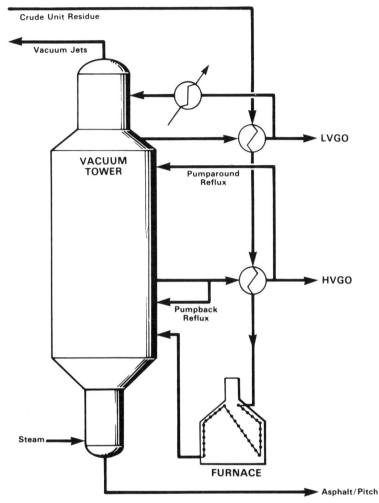

Figure 4–24. A typical vacuum-flashing unit.

LIGHT ENDS RECOVERY

Light ends recovery units separate light hydrocarbon components from the heavier continuous-boiling fractions. They also fractionate the discrete light ends components and separate the heavier continuous-boiling materials into two or more fractions. *Light ends* generally refers to hydrocarbons lighter than heptane (seven carbon atoms), thereby including everything from hydrogen through the hexanes. Configurations of light ends recovery units vary to suit the processing and economic requirements of a particular refinery.

COMMON TERMS

- **Absorption tower (absorber)** A tower which causes a separation between two components, recovering the desired amount of the heavier as part of the bottoms liquid and rejecting the lighter of the two to the overhead gas stream. A relatively heavy oil (lean oil), introduced near the top of the tower, washes back heavy components in the upflowing gas stream by the mechanism of absorption, rather than by condensing and refluxing a part of the tower overhead stream as in distillation.

- **Bubble point** The temperature at which a gas is saturated with respect to a condensable component.

- **Heavy key component** The lightest component whose percent recovery is greater in the bottoms than in the distillate (overhead stream); e.g., isobutane from a depropanizer.

- **Isomer** A molecule having the same number and kind of atoms as another, but differing from it with respect to atomic arrangement. For example:

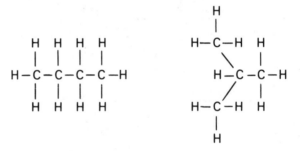

- **Lean oil** Absorption oil usually heavier than any of the components in the tower feed, or at least heavier than the light key. See *absorption tower.*

- **Light key component** The heaviest component whose percent recovery is greater in the distillate than in the bottoms; e.g., propane from a depropanizer.

- **Reflux** The part of the distillate returned to the distillation tower (column) to assist in composition control of the overhead vapors (distillate).

- **Reflux ratio** The quantity of reflux per unit to the quantity of distillate removed from the process as a product. In design, it is the ratio of liquid reflux to vapor (L/V) at any given point in a fractionating tower.

- **Saturated hydrocarbons** Hydrocarbons with all carbon bonds filled (i.e., there are no double or triple bonds as in olefins). The straight-chain paraffins like propane are typical saturated hydrocarbons.

- **Splitter** A fractionating tower that produces only overhead and bottom streams, as in the separation of an isomer from a normal paraffin (iC_4 from nC_4).

- **Stabilizer** A fractionating tower that removes light hydrocarbons from an oil to reduce vapor pressure, such as butanes and lighter components from reformate (catalytic reforming).

- **Stripper** Generally a simple distillation tower used to remove vapors from liquids, such as H_2S from light oil.

- **Unsaturated hydrocarbons** One of a class of hydrocarbons that have at least one double or triple carbon-carbon bond that is not in an aromatic ring, such as ethylene and acetylene.

LIGHT ENDS FRACTIONATION

Light ends recovery units are often found in several locations within a refinery. Their design is dictated by the processing scheme of the refinery and the type of hydrocarbon streams to be processed. Frequently, a light ends is designed to process feeds from several plants.

Refinery light ends units are described as two types: a *saturates gas plant* and an *unsaturates gas plant*. A saturates gas plant processes saturated hydrocarbons, the products from which are used in gasoline blending as feedstocks and as fuels. Its feedstock sources include crude distillation units, hydro-treaters, isomerizers, and catalytic reformers. The less common unsaturates gas plant processes unsaturated hydrocarbons from catalytic cracking units and other thermal decomposition units, the products from which provide feedstocks for petrochemical complexes.

Typically, the light ends recovery units are recognized by their series of distillation towers (see Figure 4–25). Among these towers there may be an absorber and a stripper to remove the lightest materials (ethane and lighter gases) from the gas plant's charging stock.

A popular configuration used to control distillation tower pressure is shown in Figure 4–26. This configuration, known as *hot vapor bypass*, may only be applied on a tower overhead yielding a total condensing vapor and employing a water cooled condenser.

The operation of the hot vapor bypass is as follows. Should the tower pressure control sense the pressure falling, the hot vapor bypass control valve opens to decrease the pressure differential between the condenser inlet and the receiver. Since the rundown outlet is liquid sealed, the decrease in pressure differential backs liquid up into the exchanger's shell, thus decreasing the surface available for condensing vapor. The decreased rate of condensation increases the tower pressure. Conversely, an increase in pressure differential reverses the process and decreases pressure.

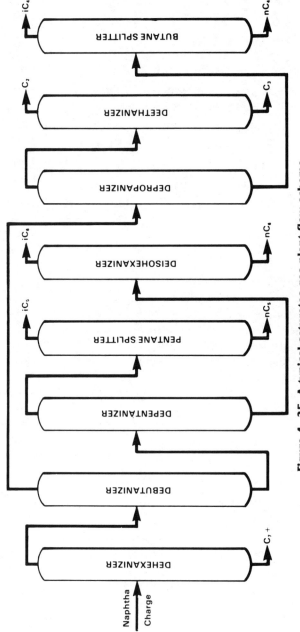

Figure 4–25. A typical saturates gas plant flow scheme.

154

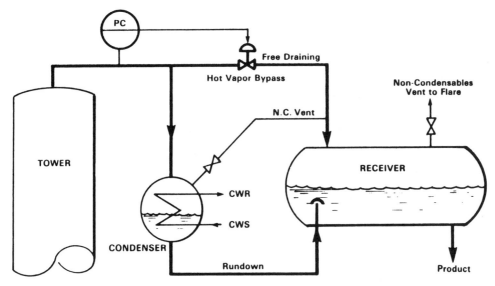

Figure 4–26. Total condensing system with hot vapor bypass.

A. *Absorber Overhead Vapor Composition.* Ordinarily, an absorber is used to separate ethane and lighter gases from a naphtha feedstock. To provide best separation, the lean oil flow rate should be directed by an on-line chromatograph measuring the minimum amount of the heavy key (propane) allowed to escape in the overhead.

B. *Stripper Bottoms Temperature.* The heat source for the boiler is generally an exchange with hot oil from a vacuum unit or another unit operating at high temperature. Consequently, deviations in oil temperature caused by upsets in the other unit will affect stripper operation. To minimize an upset, the reboiler hot oil flow rate should be manipulated as a result of a heat transfer calculation, thus enabling a constant heat input to the stripper. The set point for the rate of heat transfer is usually provided by an operator, a vapor flow rate controller or, optimally, by direction of an online analyzer.

C. *Tower Overhead Vapor Composition.* Tower reflux provides the fine adjustment for overhead vapor composition (temperature) control. Foul weather, daily temperature changes, and other ambient temperature effects will hinder the efficiency of the reflux. To combat these disturbance variables, the reflux must be compensated to handle their effects.

An online chromatograph measuring the composition of the overhead stream should adjust reflux flow to insure consistent product quality. Refer to the top portion of the tower illustrated in Figure 4–27 for details.

CONTROL SYSTEM

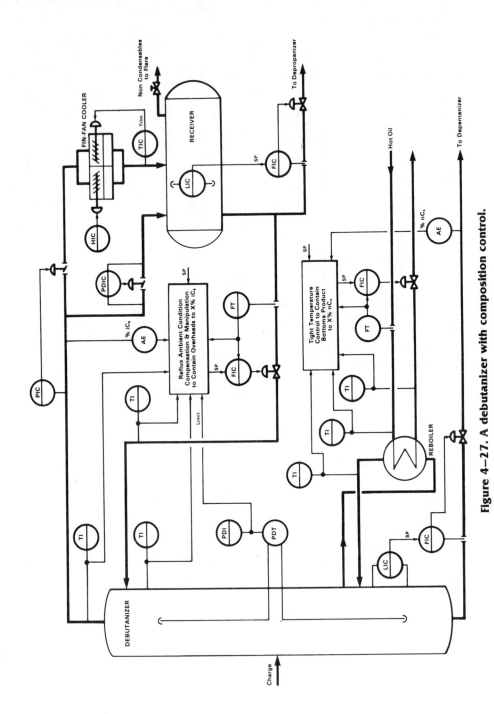

Figure 4–27. A debutanizer with composition control.

D. *Tower Bottoms Composition.* As discussed under the paragraph on *Stripper Bottoms Temperature*, the reboilers frequently receive their heat from hot oil in other refinery units. Therefore, the hot oil flow through the reboiler should be governed by a continuous heat transfer calculation to maintain a constant heat input to the tower.

Again referring to Figure 4–27, an online chromatograph measuring bottoms composition may be used to contain the amount of light key escaping to the desired minimum.

E. *Feedforward.* Feedforward manipulation of both reflux and heat transfer rates to compensate for charge rate alterations can reduce disturbance to the tower operation.

BASIC PIPELINE OPERATIONS

A pipeline is a transportation medium in competition with other common carriers. It is used to transport liquids, gases, and solids in slurry form. The most common material transported by pipelines is petroleum, as well as petroleum derivatives. Pipelines also carry fresh water to dry or remote regions and move fuel, such as coal slurries, to a generating station. This discussion will be concerned only with the transportation of liquid petroleum products by pipelines.

Petroleum pipelines fall into three major categories, named for the materials transported:

- *Crude lines* carry crude oil from various producing areas to a refinery.
- *Product lines* carry the product output from refineries or gasoline plants to the product distributor.
- *Natural gas lines* carry dry gas, usually methane, from producing areas to the consumers.

This section is mainly concerned with liquid pipelines, since crude and product lines are quite similar in operation.

In most pipelines, the product moves through the ground at approximately 1½ to 3 miles per hour (2.5 to 5 km/h). Some lines with high volume get 6 to 7 miles an hour (9.5 to 11 km/h) of product flow.

In the beginning, pipelines were developed to move crude oil from gathering areas to refineries. Now, pipelines are used for handling not only the raw material, but many types of finished products. Pipelines handle crude oils, motor oils, diesel fuels, home grade gasolines, low test and high test gasolines, jet fuel, propane, butane and ethane. It is not unusual to find the product lines handling as

PRODUCT

many as ten different types and grades of products, in 30,000 to 500,000 barrel batches. A batch in a pipeline operation is one continuous stream of the same grade and type of product isolated from other products in the line. The second product injected into the pipeline is inserted directly behind the first product. Mechanical separations, called *spheres*, are sometimes placed between the two products, but these must be removed before each pumping station and placed afterwards. In the interim, there is mixing of the two products at the station. It depends upon the philosophy of the company involved, but those that do not use the spheres for separation say that a couple hundred gallons of mixing does not matter out of a couple hundred thousand gallons. The companies that use the spheres for separation still get mixing at the stations. It comes down to a question of how much mixing of the two products can be tolerated by the customer.

In a product pipeline with batching of the products, the suction and discharge pressure set points have to be changed when a new product with a different specific gravity enters that station.

PRESSURE LOSS AND PRESSURE CONTOURS

Figure 4–28 shows a simple crude oil pipeline. Pumping stations build up the discharge pressure to make up for frictional losses in the pipe. If the pipe could be built strong enough, all pumping could be done at the input end. This would result in a line pressure contour as shown in Figure 4–29. However, the excessive cost of installing high pressure pipe prohibits that solution. Instead, pumping is done at several points, thus allowing the use of lower pressure (and less expensive) pipe.

The following is Darcy's equation for pressure drop as approximated for commercial steel pipe on a level surface.

$$P = \frac{(71.4) \ (f) \ (Q^2) \ (SG) \ (L)}{d^5} \qquad P = \frac{(2.252) \ (f) \ (p) \ (Q^2) \ (L)}{d^5}$$

where:
P = pressure psig (pascals)
f = friction factor
Q = flow rate in gallons per minute (liters per minute)
SG = specific gravity
d = inside pipe diameter in inches (millimeters)
L = length of pipe in miles (meters)
p = weight density of fluid, kilograms per cubic meter

As an example, a 300 mile crude oil pipeline with characteristics of f = 0.0204, Q = 5,000 GPM, SG = 0.88 and d = 20 inches develops a 10 psig per mile friction loss. If a single pumping station were used, the discharge pressure, shown in Figure 4–29, of 3,000 psig (20 700 kPag) would be needed. This would require a large investment per pipe, fillings, valves, etc.

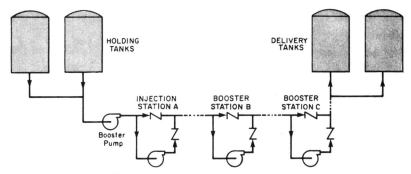

Figure 4–28. Simple liquid pipeline.

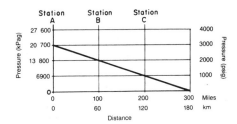

Figure 4–29. Line pressure contour.

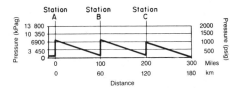

Figure 4–30. Multiple pumping station at 0 psig suction pressure.

A reasonable pipe rupture pressure might be 1,500 psig (10 350 kPa). Under this condition, the design pressure would be 1,000 psig (6 900 kPa). Figure 4–30 shows that the design pressure of the line is not exceeded, and we can still generate the same maximum flow rate as shown in Figure 4–29.

The problem with the pressure contour shown in Figure 4–30 is that it requires each station to draw its suction pressure down to 0 psig. Pump seals are very sensitive to pressure reversals caused by decreasing the inlet pressure below atmospheric pressure (or the vapor pressure of the transported fluid). Therefore, pumping at 9 psig (62 kPag) suction pressure is considered a poor practice. Secondly, some products vaporize at low pressure, causing pump erosion by cavitation and many other difficulties. To prevent this from occurring, a net positive suction head (NPSH) is held on each pump as shown in Figure 4–31. If 50 psig (345 kPag) is used for

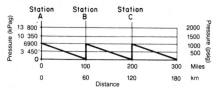

Figure 4–31. Multiple pumping station at 50 psig suction pressure.

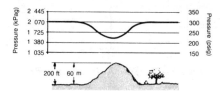

Figure 4–32. Typical pressure contour of hill.

the suction pressure, then the suction pressure at Station A is achieved by use of a high volume, low pressure booster pump. This pump takes the product from the tank and injects it into the pipeline. Also, as shown in Figure 4–31, the last station need only boost the pressure by 950 psig (6 550 kPag) because it does not have to maintain a 50 psig (345 kPag) NPSH as it is pumping into a tank at atmospheric pressure.

Figure 4–32 shows approximately how this pressure contour at the hill would look. It should be noted that a large hill could reduce the line below atmospheric pressure. This would be a place to watch for a leak (creating air pockets in the pipe) or a pipe collapse.

Another area of concern is a large valley shortly after a pumping station where the pressure in the pipe could easily increase to the rupture point. If the pressure at the top of the hill is 1,000 psig (6 900 kPag) and the hill represents a 500 psig (3 450 kPag) static head, the pressure at the bottom of the hill is 1,500 psig (10 400 kPag). For this reason, pumping stations are usually located in valleys. If this is not possible, the pipe in the valley can be protected by pressure limiting devices or systems.

PIPELINE OPERATION

Once the pipe is in the ground and the pumping equipment installed, the major variable is pumping efficiency. All pumps, motors and engines are most efficient when run at maximum design speed. Where constant speed electric pumps are used, a control valve is in the discharge, and the most efficient way to operate that station is to maintain the valve fully open. If, due to a lack of product or available tankage, the line must slow down, the operators will shut down some of the stations or shut off one or two of the pumps at each station, thereby reducing the pumping pressure. They will shut off as many as needed to reduce the flow to the required rate. They will not throttle valves to accomplish the same effect. If a flow rate is required that is between the rate developed by two pumps, then one pump is run for a percentage of the time to achieve the required average flow.

Since the product in a pipeline represents a large mass which is difficult to accelerate or decelerate, changes in flow are made as seldom as possible. These changes are usually ordered by telephone from a central office to a manned pumping station or by a completely remote, telemetered, supervisory control system. After very little operating experience, the pumping configuration can be accurately set for any required flow rate. Throttling control of the various station parameters fits into this operation only as limiting devices to protect the line against miscalculations, accidental maloperations, equipment malfunctions, or storm damage. At all other times, it is hoped that these devices will in no other way throttle the station flow.

The next problem after regulating product flow is to determine the nature of a change which can require throttling control and how the controls respond to the change. Stable control of any particular station is set up by adjusting the controller's responses with the station on the line. The loop through the throttling valve (or variable speed pump) to the pressure transmitter and controller and back to the final control element is the loop which is stabilized by the controller's response adjustments. Variations in the line flow rate change loop gain are accomplished by changing the throttling device gain. Therefore, the control responses must be set up when the loop, exclusive of the controller, is at its highest gain condition. This is usually when the line flow rate is at a minimum operating value.

STATION OPERATION

The accepted recovery curve for a station after an upset is no overshoot. Pre-Act (derivative) is not used in the controller because the pump generates a lot of noise, both in pressure and flow measurements. Transmitter line lags can be held to a minimum by careful location of the transmitters.

For pipe protection, discharge pressure is usually controlled and is further backed up by a pressure switch set at the maximum limit to shut the station down. The control set point is usually within a few percent of the switch setting, which emphasizes the need for no overshoot on the recovery curve.

As a pump and product protection device, suction pressure is also controlled and backed up with another pressure switch set close to the controller's set point. The suction and discharge controller's outputs are fed into a signal selector. Its output goes to a final control element. The selector has external feedback to the controllers to prevent reset windup.

Line flow measurement is used only as an alarm or shutdown in case of pipe breakage and as a method of determining line operation by the operators. Final delivered volume and rates are determined by tank level, either manually or by the float type telemetering transmitters.

If electric motors are used at the station, the power consumption of the motors is measured and used in the station's control system to throttle the flow when the motor becomes overloaded. This prevents complete station shutdowns due to motor overloads. In these systems, a motor overload is another override to the suction discharge control.

PIPELINE BOOSTER STATION CONTROL

As previously described in the section *Basic Pipeline Operations*, booster pumping stations are essential for maintaining sufficient pressures and flow rates in order to operate a pipeline efficiently.

Figure 4–33 shows a simplified layout of a typical booster pumping station. Booster stations increase pipeline pressure sufficiently to maintain the flow rate necessary to meet contract requirements. Flow rates are manipulated by starting or stopping pumps and stations rather than by throttling a control valve.

Station controls vary according to the operations performed and the types of equipment employed. For instance, originating stations often have pumps with flat head curves and require flow control most of the time, but booster stations are usually designed to operate at design speed. If the station consists of constant speed electric motors driving centrifugal pumps, control is exercised by throttling a control valve in the station discharge line. If the station has diesel

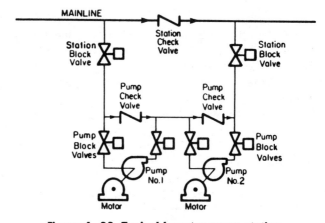

Figure 4–33. Typical booster pump station.

engines with positive displacement pumps, the final control element is a speed governor. (Variations on the control systems required by speed governors are discussed at the end of this section.)

Station control systems are designed to perform two separate functions: safety limiting and startup/shutdown control. If the basic control system cannot maintain the pipeline within the set safety limits, then automatic safety shutdown occurs.

Care must be taken in the design of station control systems to avoid spurious shutdowns. If the line is operating at capacity, one shutdown can cost a pipeline company more than the original cost of the station control package.

CONTROL SYSTEM

Safety Limits

The control system must maintain the pressure in a pipeline below its maximum design limit. If a downstream station loses power or if some malfunction in valving occurs, pressure surges can rise to dangerous levels. The control valve in the station discharge line can be throttled to maintain safe pressure. The suction pressure at a pipeline station must be maintained above a certain value to avoid cavitation in the pump. The safe suction pressure is determined by the material being moved. For instance, the safe suction pressure for propane may be 250 psi (1 700 kPa) while that for motor gasoline may be 50 psi (350 kPa). Since there can only be one control valve, the station control must be an override system. The most common control system for a pipeline station is suction-discharge pressure override control as shown in Figure 4–34.

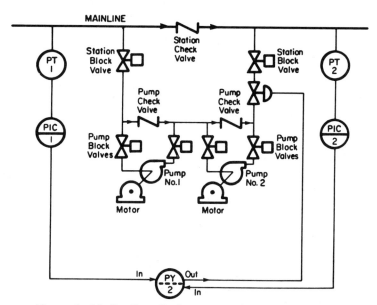

Figure 4–34. Suction-discharge override control system.

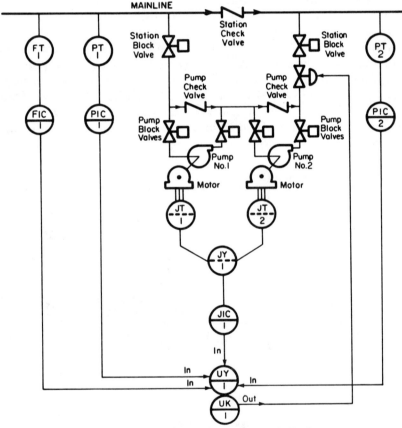

Other station parameters may require safety limit control. For example, to avoid excessive power rates based on peak demand, the power to each motor may be measured and controlled by throttling the station valve. As the flow through a centrifugal pump increases, the horsepower required (and therefore the motor current needed) increases. When the interface between two different API gravity products passes through a station, the flow rate may increase drastically. If these high flow rates are maintained, the motor can burn out. The flow rate may be measured and controlled by throttling the station control valve.

Usually a signal selector is combined with an automatic-to-manual transfer station so that the override controls are removed from the circuit when the operator selects manual control. This system is shown in Figure 4–35.

Figure 4–35. Flow, suction pressure, K.W., and discharge pressure override system.

The problems in starting a pipeline station or in starting an additional pump at the station result from the inertia of the large material mass being moved. The motor and pump can arrive at full speed much more rapidly than the total pipeline can charge to new pressure and flow conditions. Unless special care is exercised, each of the safety limits previously discussed is in danger of being exceeded. Depending on how the rest of the pipeline is operating when this station is started up, there could be an overloaded motor, a cavitating pump, or a shock wave of high pressure transmitted down the pipeline.

Startup

As an example of a booster station startup, assume the following conditions exist before starting the pump:

Manned Operation

* A suction-discharge override control system as shown in Figure 4–34 is in place. With the set point of the suction pressure controller at 40 psi (275 kPa), the actual suction pressure at 100 psi (690 kPa), and the controller in direct action.

  Discharge pressure set point is at 500 psi (3 450 kPa), the actual discharge pressure is at 100 psi (690 kPa), and the controller is in reverse action.

* The signal selector is a low current selector.

  Before starting the first pump, the operator will switch the *auto/ manual* to manual and close the control valve. The pump is started and it increases pressure in the line between the pump and the control valve (called the *case pressure*). When the pump is up to speed, the operator slowly opens the control valve while observing both discharge pressure and suction pressure. When the valve is fully open, the operator returns the controls to automatic.

Ramp Function Generator. With a device known as a ramp function generator, the station can be started remotely. The ramp function generator responds to contact closures to increase or decrease its output from 0 to 100 percent at preset rates.

Remote Operation

The ramp function generator's output will be a third input to the low signal selector. Figure 4–36 shows a control system employing the ramp function generator.

The following sequence starts the station. A contact closure activates the ramp function generator to drive its output to close the valve. The pump is then started (it usually takes five to ten seconds to get an electric motor and centrifugal pump up to speed). Another contact actuates the ramp function generator to ramp up its output.

The rate of ramp is determined and set such that the station will be brought on as rapidly as possible without exceeding any of

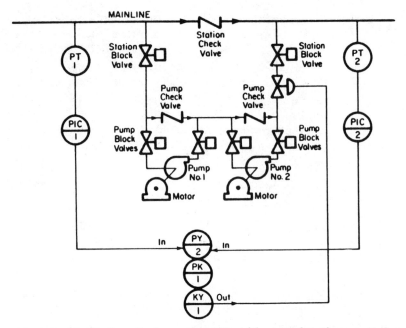

Figure 4–36. Suction-discharge override with ramp function generator.

the safety override limits. If an upset should occur during the sequence of startup, the override controllers would maintain the safety limits.

A slightly different sequence is required when one pump is operating and the second pump needs to be brought on-line. The ramp function generator is started by a contact closure but ramps down to a predetermined point. This lower limit of the down ramp is dependent upon a number of variables, such as flow rate, the product, the valve characteristics, and the size of the pumps. As a result of various products being pumped through a pipeline, the predetermined limit must be safe for all conditions.

Figure 4–37 is a graph of a ramp signal and discharge pressure during a startup using a ramp function generator. Note that the ramp rate must be set at the low rate required at first opening the valve. Also, a ramp limit is chosen to be on the safe side. Thus, optimum operating conditions must be sacrificed for safety.

Ramp Set Point Controller. With a ramp set point controller, a booster station or pump can be brought on-line in the shortest safe time with a minimum of valve throttling. The most critical of station parameters (flow, suction pressure or discharge pressure) is selected for control.

That controller then brings the station on-line at a controlled

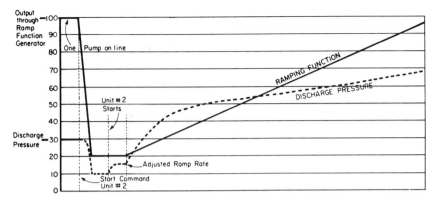

Figure 4–37. Ramp function generator startup.

ramp rate. Figure 4–38 illustrates a control system with a ramp set point controller.

Assuming the same initial conditions of suction pressure, discharge pressure, and output to the valve as before and assuming that the critical parameter for the ramp set point controller is discharge pressure, the following startup sequence is performed.

• The station receives the signal to initiate the startup sequence. As part of this sequence, a contact is provided to the float terminals of the ramp set point controller.

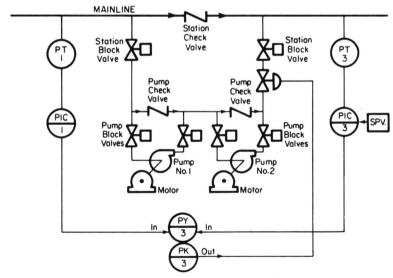

Figure 4–38. Suction-discharge pressure override with ramp set point controller.

- In the float condition, the discharge pressure controller's set point is driven at a ten-second ramp rate to the discharge pressure value.

- After ten seconds, a second contact closure is applied to the lock terminals of the set point controller. As long as this contact closure is maintained, the set point remains locked at this pre-startup value.

- The pump startup sequence is initiated. This sequence includes opening the suction valve, opening the discharge valve and monitoring the pump parameters.

- As the discharge pressure starts to increase, the controller immediately starts to drive the control valve closed to prevent the discharge pressure from increasing beyond the discharge pressure set point.

- When the pump is up to speed, all the additional pressure developed by the pump is dropped across the control valve. Therefore, the discharge pressure of the station and the flow past the station have not increased.

- Now the contact closures (float and lock) are removed, and the set point ramps at its preset rate to the remote set point value. The controller will send whatever signal is necessary to the valve to control the pressure at the required discharge pressure set point.

The sequence works equally well in bringing a booster station on-line or bringing a second pump on-line with the first pump running. Figure 4–39 is a graphical representation of this advanced method of station or pump startup.

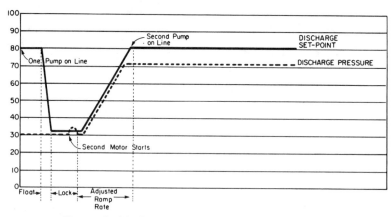

Figure 4–39. Ramp set point controller startup.

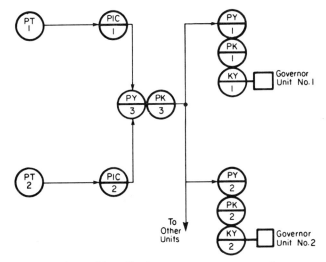

Figure 4–40. Positive displacement pump control system.

Positive Displacement Pumps. Most positive discharge pumps are diesel driven and variable speed. Therefore, the final control element is a speed governor. The control system is similar to the previous systems, except that the selector output goes to the speed governors of each pump, instead of going to a control valve. A ramp function generator (R.F.G.) is added for each unit to bring them up to speed. An example of this system is shown in Figure 4–40.

INSTRUMENT LIST

	Tag No.	Description
Suction-discharge override control	PT 1	Suction pressure transmitter
	PT 2	Discharge pressure transmitter
	PY 2	Signal selector
	PIC 1	Suction pressure controller
	PIC 2	Discharge pressure controller
Flow, suction pressure, K.W., and discharge pressure override	PT 1	Suction pressure transmitter
	PT 2	Discharge pressure transmitter
	FT 1	Flow transmitter
	JT 1	Kilowatt transmitter
	JT 2	Kilowatt transmitter
	JY 1	Signal selector
	UY 1	A/M station with selector
	UK 1	Limiter
	PIC 1	Suction pressure controller
	PIC 2	Discharge pressure controller
	FIC 1	Flow controller
	JIC 1	Motor load controller

	Tag No.	Description
Suction-discharge override with ramp function generator	PT 1	Suction pressure transmitter
	PT 2	Discharge pressure transmitter
	PY 1	A/M station with selector
	PK 1	Limiter and ramp function
	KY 1	Generator
Suction-discharge override with ramp set point controller	PT 1	Suction pressure transmitter
	PT 3	Discharge pressure transmitter
	PY 3	A/M station with selector
	PK 3	Limiter
	PIC 1	Suction pressure controller
	PIC 3	Discharge pressure controller, ramp set point
Positive displacement pump control	PT 1	Suction pressure transmitter
	PT 2	Discharge pressure transmitter
	PY 3	A/M station with selector
	PK 3	Limiter
	PY 1	A/M station with selector
	PK 1	Limiter and ramp function
	KY 1	Generator
	PY 2	A/M station with selector
	PK 2	Limiter and ramp function
	KY 2	Generator
	PIC 1	Suction pressure controller
	PIC 2	Discharge pressure controller

5 Pulp and Paper

WOOD CHIP HANDLING

The wood chips, arriving at the paper mill, are stored in piles outside the mill (see Figure 5–1). One pile contains low density chips and the other high density chips.

Below each pile is an endless screw conveyor which supplies chips to the belt conveyor. These screw conveyors not only rotate but also move horizontally along the length of the pile. This involves the control of two drive motors per screw conveyor. From the screw conveyors the chips land on a common belt conveyor that carries them to the disk screen.

The endless screws are driven by variable speed 150 HP DC motors which are regulated by a level controller at the chip bin. A differential value between the speed of the two screws is due to a greater requirement of one type of chip over the other type.

The chips from both piles fall onto the same conveyor belt that carries them to the disk screen. Before the chips reach the disk screen, they pass under an electromagnet suspended above the belt conveyor that removes most metallic objects like nuts, bolts or any machine parts from entering the mill with the wood chips.

The disk screen is used to remove large blocks of frozen chips, rocks, etc. in order to protect the pneumatic conveyor and other process machinery. In the disk screen bin, there is a high level detector that will stop the screws and the conveyor belt to prevent bin overflow.

From the disk screen bin, the chips are pneumatically transported to a cyclone in the mill. This is accomplished using a large compressor and a special feeder that prevents pressure losses at the input of the pneumatic conveyor. The feeder has a sealed rotary valve used to transfer the chips to the cyclone. From the disk screen bin, the chips are blown into a cyclone where the excess air pressure is relieved so that the chips do not enter the chip screen under pressure.

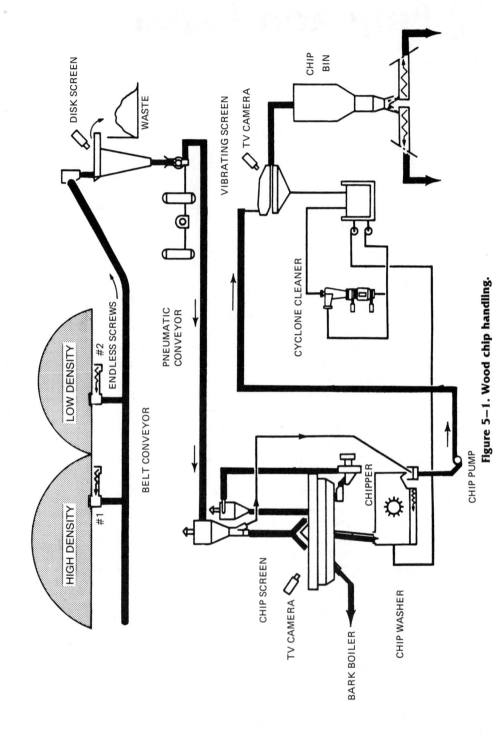

Figure 5—1. Wood chip handling.

172

Once the chips have reached the chip screen, they are sorted into three sizes (see Figure 5–2). Since there is an ideal chip size in the refining stage, the chip screen is used to eliminate those that are too large or too small. The ideal chip size is approximately ⅝ inch to ⅞ inch (16 to 22 mm) and smaller or larger ones should not be used. The larger rejected chips are sent to a chipper where they are cut to an acceptable size and sent back to the chip screen. The smaller chips and fines are either discarded or burned in the mill's bark boiler.

From the chip screen the chips fall into the chip washer (see Figure 5–3). At this stage foreign material is removed: any sand, rock, gravel or metals missed by the electromagnet on the belt conveyor are eliminated. The foreign matter sinks to the washer bottom and is picked up by a screw conveyor which forces the waste out of the washer. It is then sent to an automatic time delay discharge unit which has a valve above and below it. These valves are never opened or closed at the same time. Their purpose is to prevent water from being discharged with the waste. The waste is removed from the mill in containers.

The clean chips are discharged to a chip vat, a holding reservoir which insures a constant flow of chips to the chip pump. Once the chips have been washed, it is possible to use a conventional pump to move the chips to the vibrating screen. Here, water from the chips can be removed before the chips go to the chip bin. The vibrating screen holds the chips while letting the water seep through and fall into a wash water reservoir. The water in this reservoir is recycled to be used again in the chip washer.

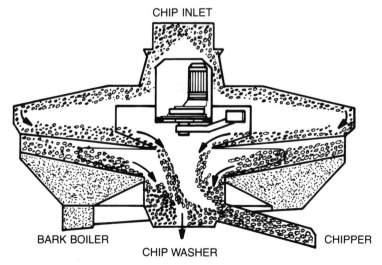

Figure 5–2. Chip screen.

From the vibrating screen, the now dried chips are transferred to the chip bin which governs the whole process. The level in the chip bin controls the endless screw conveyors at the chip piles. From this bin the chips are supplied to the refiners. Below the bin there can be multiple conveyor screws that feed refiner stages.

CONTROL SYSTEMS

Endless Screw Conveyor

The control system of the screw conveyors at the chip piles involves the control of the rotation speed of the screws (see Figure 5–4).

There must be an accurate mixture of high and low density wood chips, depending on the desired paper quality. This is accomplished using ratio modules on the output signal of the chip bin level detector controller.

The chip bin is equipped with a level control system to maintain a constant level in the bin. The output signal of the controller goes to the ratio stations. Selected at the ratio stations is the correct mixture of the two types of chips. The two outputs from the ratio stations are sent to their corresponding endless screw conveyors. The signals received are used to adjust the speed of the screws.

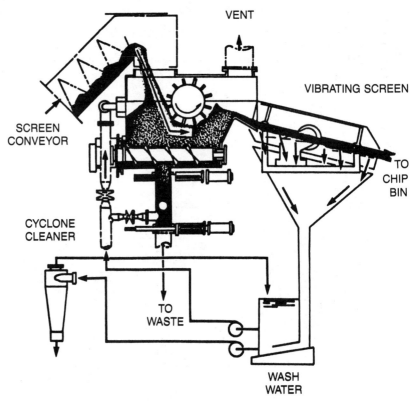

Figure 5–3. Chip washer.

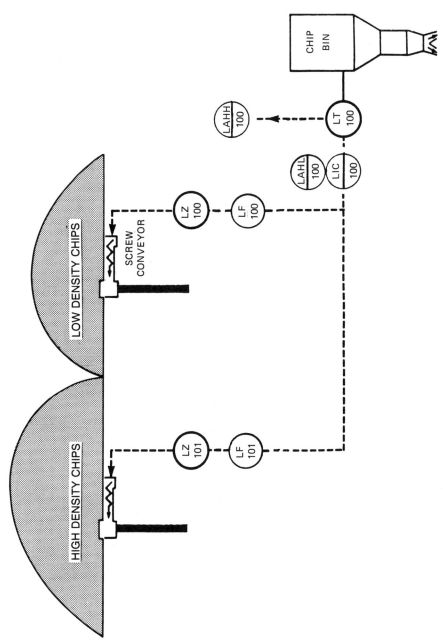

Figure 5–4. Endless screw conveyor control system.

To this system is added an override protection circuit which is governed by the disk screen bin. When the bin reaches maximum level, a signal is sent to cut off the feed of chips and both screws are stopped as well as the belt conveyor.

An often omitted control system section is the use of closed circuit television monitoring system. There should be a camera installed at the disk screen such that both the disks and the belt conveyor leading to it can be seen clearly. The purpose of this is to view the amount of material entering the screen. This can determine if there is a screw malfunction preventing the wood chips from getting to the belt conveyor. In addition to the above location, a camera should be located at the chip and vibrating screens. These places are where a problem will occur most often that will not be detected by measuring instruments.

Chip Bin   Instrumentation at the chip bin consists of a level and a temperature control system (see Figure 5–5). Level control in the bin regulates the speed of the two endless screws at the wood chip piles. The wood chip level in the bin is measured using a weak radioactive source (gamma rays) and a radiation detector which gives accurate results. The output signal from the controller is linked at the ratio stations, which send it to the individual screw conveyors. Also found in this loop are alarms. There are low and high alarms to warn operators and one critical high alarm which stops all conveyors preceding the disk screen. The temperature control system is used to maintain a constant temperature in the chip bin of about 90°F (32°C).

INSTRUMENT LIST

	Tag No.	Description
Endless screw	LT 100	Chip level transmitter
conveyor control	LAHH 100	Critical high chip level alarm
system	LIC 100	Chip level controller
	LALH 100	High and low chip level alarms
	LF 100	Low density chip ratio station
	LZ 100	Low density chip screw conveyor speed control
	LF 101	High density chip ratio station
	LZ 101	High density chip screw conveyor speed control
Chip bin control	LT 100A	Chip level transmitter, detector
system	LT 100B	Chip level transmitter, source
	TT 100	Chip bin temperature transmitter
	TIC 100	Chip bin temperature controller
	TAHL 100	High and low chip bin temperature alarms
	TCV 100	Temperature control steam valve

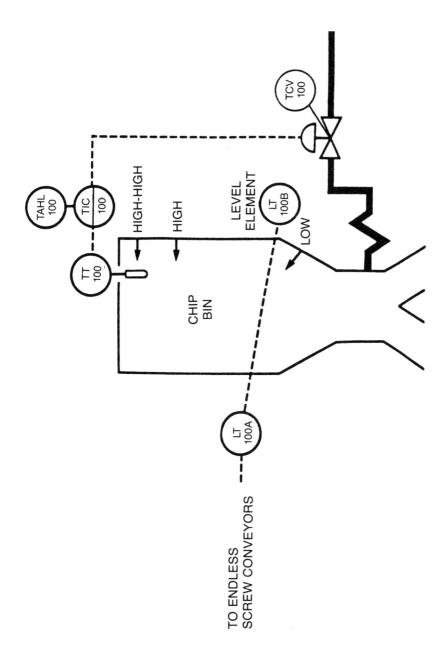

Figure 5–5. Chip bin control system.

177

SULFATE CHEMICAL DIGESTER

Paper mills produce products which range from fine papers to corrugated cardboard. This range of products requires that the basic raw material of paper (wood pulp) has the fiber characteristics to meet the specific grade manufacturing requirements. Numerous pulping processes have been developed to produce economically the exact pulp required. These processes include mechanical and chemical means of digestion and combinations of both. They can be run as batch or continuous processes.

This section will discuss the control systems used in the sulfate, kraft, chemical pulping batch type digestion process.

THE CHEMICAL DIGESTION PROCESS

In the chemical digestion process, wood chips are cooked under controlled temperature and pressure conditions to dissolve the fiber bonding agents and other nonfibrous portions of the wood chip. The resultant residue is primarily a cellulose fiber mixture called *pulp*.

MAJOR PROCESS EQUIPMENT COMPONENTS

The wood chips and chemical cooking liquor are charged into a large vertically positioned pressure vessel called the *digester main shell* (this can be seen in Figure 5–6). Chips are added on a weight basis while liquor is metered in on a volumetric basis. The automatic capping valve closes over the digester chip charging port, this initiating the cooking cycle. The liquor circulation pump then begins extracting the cooking liquor through the chip screens. The liquor is then pumped through a tube-in-shell heat exchanger, called the *liquor heater*, and back into the digester. As the cook progresses, air and other noncondensables are vented through the digester relief line. Upon completion of the cook the pulp is discharged through the blow valve which seals the bottom connected blowline.

COOKING METHODS

The two methods of cooking in common use are indirect steaming and direct steaming.

Indirect Steaming

In the indirect method, shown in Figure 5–6, the cooking liquor is pumped from the digester, passed through a heat exchanger and returned to the digester.

This return flow is adjusted to maintain a larger flow in the top circulation line in order to compensate for the liquid head on the chips in the lower portion of the digester. This liquid head contributes to more rapid cooking near the bottom of the digester. Therefore, the larger flow of heated cooking liquor is maintained in the top of the digester to balance the overall digester cooking rate.

Indirect steaming systems are used where a greater degree of control on pulp quality is required. This control is essential where

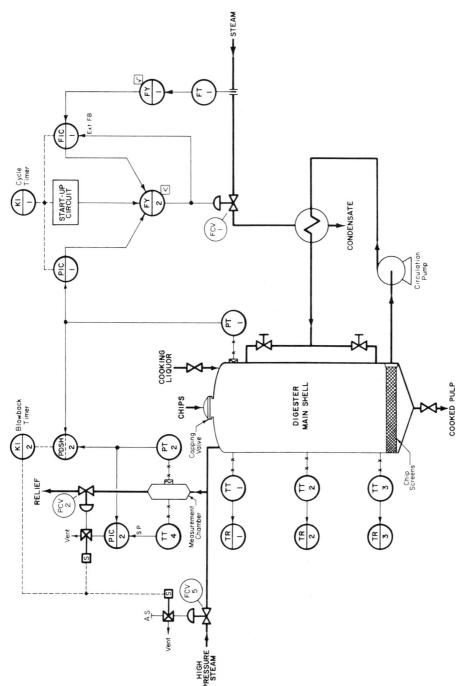

Figure 5–6. Indirect steaming method of sulfate batch digester control systems.

179

the pulp must undergo a subsequent bleaching process or where varying grades of market pulp are being produced to specific customer requirements.

Direct Steaming

In the direct steaming method, the steam is directed into the bottom of the digester and comes into direct contact with the chips. This method is used when the important parameter is high volume pulp production per digester. Short batch cooking times, due to the direct injection of the steam into the chip mass, makes greater production possible. Pulp quality is poorer than with the indirect steaming method. The chip mass is prevented from attaining a uniform cooking condition due to circulation and heat transfer problems associated with the direct steam injection.

BATCH DIGESTERS

The batch digester is chosen over the continuous type for the following applications:

- Batch digesters are used mainly in multigrade mills where production of many grades of paper requires a wide range of pulps.

- Batch digesters require less maintenance than continuous digester. Since it is common to produce all one pulp grade in one continuous digester, a breakdown in any major component of the digester can stop the entire pulp production of that grade. Where batch digesters are used, more than one batch digester is usually available, allowing a redistribution of the production loss.

CONTROL SYSTEM

This batch digester control system may be applied to either direct or indirect steamed digesters utilizing the sulfate process. The design goal of a batch digester control system is to maintain the critical parameters of time, temperature and pressure throughout the cooking cycle. Flexibility adjustment of the indices of these parameters is also a design objective.

Steaming System

The steaming control system consists of steam flow and digester pressure feedback control loops. The digester pressure control loop functions as an override for the steam flow control loop.

Historically, batch digester control has been implemented by utilizing cam program controllers for programming digester pressure. Today, these are largely replaced by electronic programmable controllers. As weak species vary or other factors change a new cam is required. In order to overcome this disadvantage, the digester control system shown in Figure 5–6 accomplishes the ramping to cooling pressure by regulating the steam flow to the digester. This controls the ramp rate to cooking pressure by fixing the BTU input at a controlled rate. Duration of cook is controlled by a timer KI 1. This allows the operator to change easily the ramping time to cooking pressure and temperature by changing the steam flow controller, FIC 1, set point. No recutting of cams is necessary and repeatable ramp times are obtained from batch to batch. There is no overpeaking upon reaching the cooking pressure as set on the pressure controller, PIC 1, set point. Upon startup, a delay circuit limits the initial steam valve opening, preventing large sudden load changes of the steam demand on the power house. After a few minutes, the limiting action is removed and full steam flow control regulates the digester steaming rate. The digester pressure begins to increase but is limited by the steaming rate set on the steam flow control. As the digester reaches cooking pressure, the digester pressure controller, PIC 1, overrides the steam flow control through signal selector FY 1. To maintain the cooking pressure in the main digester shell, pressure is chosen as the major cooking control parameter, representing as it does an overall measurement of digester temperature at the corresponding saturated steam pressure condition. True representative temperatures, based upon circulating cooking liquor or temperature within the chip mass of the chip charge, are difficult to obtain. The time of the cook at the cooking pressure is adjustable and easily changed if required. Upon completion of the cook, the system automatically shuts down and initiates a cycle complete alarm. The operator can then empty (blow) the digester into the blow tank.

The purpose of the relief control system is to vent entrained air and other noncondensables given off from the chip mass during the cooking cycle.   **Relief System**

The air and other noncondensable vapors create a partial vapor pressure above that of saturated steam at the cooking temperature. Since total pressure is controlled by PIC 1, the existence of a partial pressure of saturated steam above that lowers the steam demand into the main digester shell.

INSTRUMENT LIST

	Tag No.	Description
Steaming control loop	KI 1	Cycler timer
		Startup circuit
	PT 1	Digester pressure transmitter
	PIC 1	Digester pressure transmitter
	FT 1	Steam flow transmitter
	FY 1	Square root extractor
	FIC 1	Steam flow controller
	FY 2	Low signal selector
	FCV 1	Digester steam flow control valve with electro-pneumatic transducer and valve positioner
Relief control loop	KI 2	Blowback timer
	PT 2	Relief line pressure transmitter

Tag No.	Description
PDSH 2	Blowback differential switch
PT 4	Relief line temperature transmitter
PIC 2	Relief line pressure controller
PCV 2	Relief line pressure control valve with electro-pneumatic transducer and valve positioner

Main shell monitoring loop

Tag No.	Description
TT 1	Top temperature transmitter
TR 1	Top temperature recorder
TT 2	Middle temperature transmitter
TT 3	Bottom temperature transmitter
TR 3	Bottom temperature recorder

COOKING LIQUOR MEASURING SYSTEMS

The sulfate (kraft) digestion process involves the cooking of wood chips immersed in caustic cooking liquor at an elevated temperature and pressure. The cooking liquor consists of the concentrated caustic liquor referred to as *white liquor* and dilute from the brown stock washing area called *weak black liquor*. The concentration of cooking liquor in the digester during the cook greatly influences the quality and quantity of useable pulp produced from any given cook. This section will discuss the batch metering type of liquor measuring used to charge precise amounts of white and black liquor into batch digesters.

Control Systems and the Liquor Measuring Process

The overall purpose of the liquor measuring system is to charge exact amounts of white and black liquor into the batch digester. The design goal is to assure that the exact concentration and quantity of cooking liquor required for the cook is present in the digester. This is accomplished by charging exact amounts of a known concentration with black liquor. Obvious benefits from precise liquor measurement are:

- Higher pulp yields
- Better overall quality control
- Lower operating costs

Batch Flowmeter System

The batch flowmeter system consists of an electrical sequencing and interlock system controlled from a flow measurement and integration loop. It may be applied to any process which requires a fixed amount of the liquid addition on a batch basis. However, the control system configuration discussed in this section specifically applies to charging cooking liquor into batch digesters. The concept of the batch metering method is based upon integration of volumetric flow. In order to implement this concept, a flow integration system consisting of a magnetic flowmeter, FE 1, integrator with pulse outputs, FQT 1, and a predetermining counter, FQIS 1, are utilized.

Both the white and black liquor measuring systems operate the same. The overall system is shown in Figure 5–7 and a typical wiring schematic is shown in Figure 5–8.

Charging System

The amount of each liquor to be charged is set on the predetermining counter FQIS 1, and the fill cycle is initiated. Providing that all interlock conditions of flow digester pressure (pressure contact PSL 1 at the normally closed position) and filling valve position status switch ZSH 1 are normal, the system is energized and begins filling the digester. The liquor pump and fill valve remain energized until coincidence is reached on the determining counter. When coincidence is reached, the counter contacts transfer closing the liquor fill valve and stopping the liquor transfer pump. This completes the liquor charge cycle until a new batch is required by the operator.

As the steam input is reduced, the cooking temperature in the digester is reduced to a point lower than that of saturated steam temperature at the corresponding total pressure. This results in incomplete cooking of the pulp in the normal cycle time, creating production delays. This system consists of a single feedback control loop with remote adjustment of the set point.

A filled thermal system temperature transmitter, TT 1, whose output is calibrated to follow the saturated steam temperature versus the pressure curve, is used as the set point for the relief line pressure controller, PIC 2. If the relief line pressure is above the saturated steam pressure equivalent to the temperature in the digester, the relief controller takes corrective action until its set point is satisfied.

Blowback System

The purpose of the blowback system is to prevent plugging of the screens on the relief line. It also minimizes pullover of cooking liquor from the digester body through the relief line. Under normal automatic operation, the relief line pressure and digester main body pressure are compared in the differential pressure switch, PDSH 2. When the differential rises from 5 to 7 psig (35 to 50 kPag), a timer, KI 2, is energized. The timer closes the relief control valve, PCV 2, and opens the high pressure steam valve, FCV 5, blowing steam back through the relief line screens.

Application Data

Digester pressure and top, middle, bottom and circulating liquor temperatures are normally recorded. These records serve as an index of product uniformity throughout the digester as well as from batch to batch.

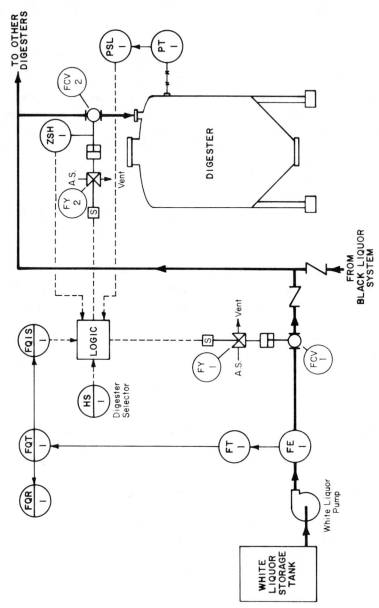

Figure 5–7. Liquor measuring system.

184

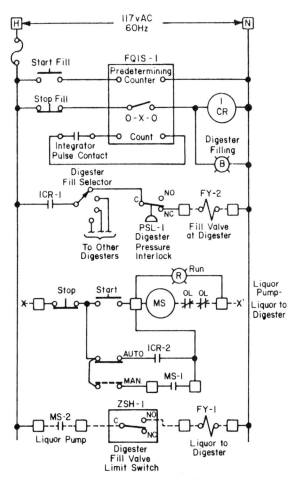

Figure 5–8. Typical wiring schematic for liquor measuring system.

INSTRUMENT LIST

Tag No.	Description
FE 1	Magnetic flowmeter
FT 1	Magnetic flowmeter—transmitter
FQT 1	Integrator—pulse output
FQIS 1	Predetermining counter
FQR 1	Printer
FCV 1	Liquor control valve to digesters
FCV 2	Liquor control valve to digester no. 1
PSL 1	Alarm switch
PT 1	Digester pressure transmitter
CR 1	Relay

DIGESTER BLOW HEAT RECOVERY

Pulp mills require large quantities of hot water in order to wash the cooking chemicals from the pulp after completion of the digestion process. Approximately 2,000,000 BTUs (2 326 000 kJ) per air dry ton of pulp produced are available for water heating purposes when a batch digester is emptied into the blow tank. In order to recover this large amount of heat energy, the steam flashed from the digester blow tank is passed through a heat recovery system. The direct contact heat recovery system is the subject of this section. Specifically, the direct contact condenser or jet type is discussed.

THE HEAT RECOVERY PROCESS

The direct contact condenser heat recovery process produces hot water on a continuous recirculation of hot condensate from the hot water accumulator through the heat exchanger (see Figure 5–9). When a digester blow occurs, the temperature switch TSH 1 senses by temperature rise the presence of the inrushing steam. Pump P 2 is energized and begins delivering spray water to the direct contact condenser. The blow steam is condensed and falls into the upper portion of the hot water accumulator. A baffle arrangement prevents intermixing of the hot condensate with the cooler water already present in lower portions of the accumulator. The hot condensate is withdrawn from the upper portion of the accumulator. It is then pumped through filters and the heat exchanger where its heat is transferred to the makeup water for the hot water storage tank. The cooled condensate then reenters the hot water accumulator where it serves as the makeup for the cooling water to the direct contact condenser. Some pulp mills transfer the condensate directly from the upper portion of the hot water accumulator to the hot water storage tank. This is more efficient in heat recovery since it involves no heat exchange. However, since the water may be contaminated from chemicals left from the digestion process, lower washing efficiency occurs and strong odors may be present in the washing area.

CONTROL SYSTEM OBJECTIVE AND CLASSIFICATION

The design goal of the batch digester blow heat recovery system is to maintain the optimum temperatures and levels throughout the system in order to recover the maximum amount of heat per digester blow. The overall control system consists of feedback control systems for condensate temperature, hot water temperature and hot water storage tank level. In addition, the hot water storage tank level functions as an override system through the high signal selector. The purpose of the override system is to prevent overflow of the hot water storage tank.

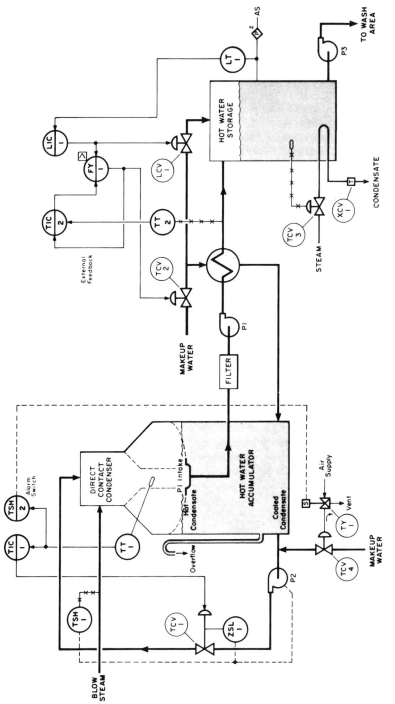

Figure 5–9. Batch digester—blow heat recovery control system.

187

The control systems discussed are universally applied for heat recovery systems, but the process and systems arrangement discussed in this section are specific to pulp mills utilizing batch digesters.

CONTROL SYSTEM

Condenser Accumulator Water Temperature System

The purpose of this control system is to insure complete condensation of the flow steam by maintaining condensate temperature at approximately 200°F (90°C). This is accomplished by throttling the condenser spray control valve, TCV 1. When a digester blow occurs, temperature switch TSH 1 senses the presence of incoming steam by a temperature rise. TSH 1 contacts close, energizing the hot water accumulator circulation pump P 2 into providing cooling water for the direct contact condenser. Condensate temperature control loop TIC 1 monitors the temperature of the hot condensate. Limit switch ZSL 1 detects the almost closed position of TCV 1 indicating completion of the digester blow inferentially as very little cooling water is being used. This shuts down the circulation pump until the next blow is signaled by TSH 1, the temperature switch.

High Temperature Alarm System

The purpose of the high temperature alarm system is to assure a supply of cooling water for the condensing of steam during startup or other conditions where the accumulator does not have an adequate cooling water supply. This is accomplished by opening TCV 4 makeup water control valve to provide spray water to the condenser. After initial startup, the accumulator will normally remain full of water since water in the form of condensing steam is being added during each blow period. Excessive water overflows and is drained after the blow heat has been removed.

Hot Water Storage Temperature System

The function of the hot water temperature control loop is to maintain the required hot water-to-storage temperature. As heat as a result of a digester blow is made available to the heat exchanger, the temperature controller TIC 2 regulates the makeup water to the heat exchanger in order to maintain the required hot water temperature.

Hot Water Storage Tank Level System

The level control loop LIC 1 assures that an adequate amount of water is always available even though there are usually large water quantity demands from other mill areas. The level control loop also functions as an override of the temperature control makeup water system to prevent overflow of the hot water storage tank. In the event of a high level in the hot water storage tank, the level controller signal is selected by the high signal selector FY 1 to close the makeup water valve.

Note that the temperature control system functions to move the heat forward to the hot water storage tank as it becomes available.

It is limited only by the heat available from the exchanger or by a high level condition in the hot water tank.

Because of the occurrence of large water demands or long shutdown times in the washing area, TCV 3, a self-acting temperature controller, maintains the hot water storage temperature by throttling the steam flow to the coil type tank heater.

Application Data

External feedback is required on the hot water-to-storage temperature controller. External feedback prevents reset windup during periods of large water usage when the hot water storage level controller output is in control. The hot water storage level controller requires only the proportional response negating any external feedback requirement.

The circulating water and the hot water-to-storage temperature are normally recorded. These records serve as an index of heat transfer efficiency and are used as maintenance guides for checking the heat exchangers, filters and pumps.

INSTRUMENT LIST

	Tag No.	Description
Hot water accumulator control system	TT 1	Condensate temperature transmitter
	TE 1	Condensate temperature element
	TIC 1	Condensate temperature controller
	TCV 1	Condensate temperature control valve
	ZSL 1	Limit switch
	TSH 1	Temperature switch
	TY 1	3-way solenoid valve
	TSH 2	High temperature alarm
	TCV 4	Makeup water control valve
Hot water storage control systems	LT 1	Level transmitter
		Purge control station:
	P	Purge air pressure reducing valve
		Purge air rotameter
	LIC 1	Level controller
	LCV 1	Makeup water control valve
	TT 2	Temperature transmitter
	TIC 2	Temperature controller
	FY 1	High signal selector
	TCV 2	Cooling water control valve
	TCV 3	Temperature controller

THERMOMECHANICAL PULPING

There are two ways of making pulp from wood chips. One method is through the use of mechanical force and the other is through chemical action. Thermomechanical pulping is a combination of thermal and mechanical energy breaking down wood chips to make pulp. The process involves passing the chips through a small gap between revolving disks to break apart individual fibers.

TMP REFINING STAGES

There are a number of thermomechanical processes. The difference in refiner types is not major but each system must be looked at on an individual basis. In this section, we will follow the wood chips to the refiners where we will give a basic explanation of the various types of refiners available (see Figure 5–10).

Chip Cooker

A conveyor screw takes the wood chips from the chip bin and feeds them to a conical screw. This screw seals the pressure from this point on in the system. It has a special hydraulic piston which maintains a plug of chips and so prevents pressure from decreasing in the system.

In the cooker, the chips are preheated to a temperature at which the lignin will soften, making it easier for the refiner to separate the fibers. The final chip temperature depends on the amount of time the chips stay in the cooker. If the chips stay too long or are heated to too high a temperature, they will "burn" and not offer a good quality of paper. The normal retention time is that of around three minutes at a cooker pressure of 7 psi (50 kPa).

From the cooker the chips are fed to the primary refiners using pressurized conveyor screws. Pulp production is dependent on the speed of these screws.

Refiner Types

There are three types of refiners in use today. One type is the refiner with one single rotating disk. Another type has two oppositely rotating disks. One which is no longer so popular consists of two rotating disks and one center stationary one. In this section we will be looking at only the single rotating disk type refiner.

As the wood chips enter the refiner, they are picked up by a fixed speed screw that feeds the refiner. This screw forces the chips through the stationary disk, between the revolving disk (1800 rpm), and out the refiner bottom. From between the disks, a centrifugal force is exerted on the chips which makes them move towards the outside of the disks and at the same time by passing between the plates the chips are separated into fibers. The substance that comes out of the primary refiner is a coarse pulp which must go through the secondary refiner in order to obtain a much finer pulp.

Two types of refiner stages are available. One type has the primary refiner pressurized and the secondary at atmosphere whereas

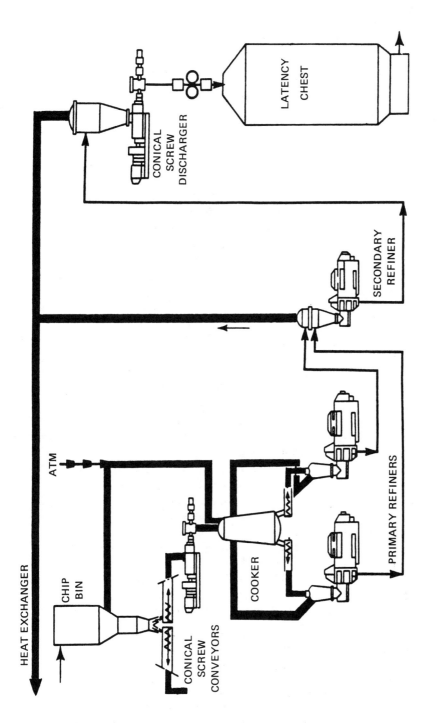

Figure 5–10. TMP refining stages.

191

the other type has both the primary and secondary refiners under pressure.

In the case where the primary refiner is the only one pressurized, we find a control system consisting of a blow valve that maintains a constant pressure in the refiner. The pressure in the refiner must be maintained at a constant level in order for the plate gap control system to work efficiently.

In the case where both refiners are pressurized, a blow valve is located after the secondary refiner. The advantage of having both refiners pressurized is that the efficiency of the system can be increased, resulting in a higher quality of pulp. Since both refiners are pressurized, the operating temperatures are increased which in turn facilitates the separation of chips into fibers. At lower temperatures this action takes a great deal more energy, meaning higher operating costs and lower pulp quality.

Refiners are large pieces of machinery and can be damaged very easily. The disks on the refiners are around 60 inches (1.5 m) in diameter and are driven by motors as powerful as 15,000 HP (11 MW) with a gap between the plates of 2/100 inch to 5/100 inch (0.5 to 1.0 mm). An independent plate protection system is generally part of the control scheme. This consists of a computer which measures plate gap, vibration level, power of the driving motor, and other variables such as oil pressure, etc. Should the computer detect offsets greater than the permissible differences, automatic corrective action immediately takes place. For example, if a too high vibration level is detected, the refiner plates are withdrawn from each other and the chip feed motor turned off.

One reason for pulp consistency change in the refiner would be a change in the flow of the dilution water entering the refiner. This is generally a control loop on a refiner which regulates a given amount of water added to the refiner to maintain a desired consistency of the pulp.

Aside from the control systems that we will discuss in the next section, there are also a number of controls on the individual refiner and its drive motor. These are oil flow, oil pressure, temperature of the refiner, and motor lubrication systems. Also monitored are pressure and flows of the water used to cool the electric motor.

REFINER STAGE CONTROL SYSTEMS

Due to the overall complexity of the control systems in the refining stages, each loop will be broken down to cover only the control elements applicable to the individual process section.

Conveyor Screws

On the chip bin output, there are three conveyor screws of which one is a conical screw and the other two are endless (see Figure 5–11). On startup, a refiner endless screw conveyor speed is selected and the set point adjusted on screw speed controller SIC

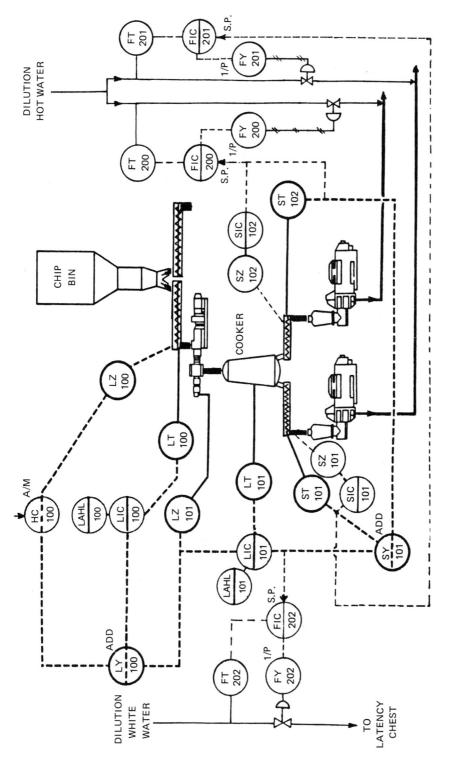

Figure 5-11. Conveyor screw control system.

193

101. The refiner screw speed is used as a set point for the cooker level controller. If there are two refiners on the output of the cooker, the outputs of the screw speed transmitters are added and the sum becomes the set point for the cooker level controller. The cooker level controller output governs the speed of the conical feed screw. In turn, two other signals are added: the speed of the conical screw and the output on the cooker bin level controller. The sum of these becomes the speed signal for the chip bin screw conveyor motor.

Aside from the control system, there are safety features included in certain loops. A low speed alarm on the pressurized screw conveyor, a high and low level alarm on the cooker, a high and low level alarm on the cooker feed bin, and a low speed alarm on the chip bin conveyor screw protect the system.

Cooker

On the cooker there is a pressure control loop that involves split range valves (see Figure 5–12). This means that one valve operates in the lower half of the control signal and the other valve in the upper half. On the cooker we find a pressure transmitter. It sends its signal to a pressure controller which operates the split range valves. When the pressure is too low, the valve on the steam line entering the cooker reacts by opening to a decreasing signal from the controller. As the pressure increases above the set point, the

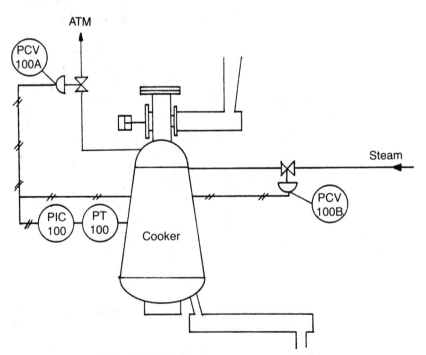

Figure 5–12. Cooker pressure control.

steam valve is closed and the exhaust valve opens up to decrease pressure. When the system is at equilibrium, both valves are closed. If the system should break down, a failsafe design insures that the steam valve would remain closed and the exhaust valve open.

Dilution water addition throughout the system is controlled by input signals set at different points from the above mentioned loops. The set point for the dilution water flow controller at the refiner screw is the output signal from the pressurized screw speed transmitter. The set point for the dilution water addition controller on the refiner pulp line just before the secondary refiner's cyclone is also the pressurized screw conveyor's speed transmitter output signal. If this signal is low, a low speed switch will open a solenoid valve that will stop the dilution water from entering the cyclone. The output signal of the adder, on the output of the two pressurized screw conveyor speed transmitters, is the set point for a water flow controller at the latency chest. This signal is also monitored by a low speed switch which controls a solenoid on the white water pipe leading to the latency chest input.

Dilution Water

On each refiner pulp output line there is a pressure transmitter whose signal is received by a pressure controller (see Figure 5–13). The pressure controller signal is used for two split range valves. As the pressure increases, the blow valve opens to allow the process to match the set point. When the pressure falls below the set point, the valve on the startup steam line opens. When the system is in operation, the steam valve is closed since the refiner itself generates steam in sufficient quantity to maintain the desired operating pressure. This steam is produced by heat generated through the refining action and water contained in the wood fibers. The blow valve controls both refiner pressure and pulp output from the refiners. To prevent output line blockage the blow valve has a mechanical stop to prevent it from closing more than twenty-five percent. The signal from the pressure transmitter is also linked to a low pressure switch which prevents the drain on the pulp line from opening during normal operation. This solenoid can also be activated by the logic controller.

Refiners

Protection on the pulp lines includes a solenoid valve that closes the line if low pressure is detected downstream. This prevents pulp out of the secondary refiner cyclone from backing up into a stopped primary refiner.

At the secondary refiner input, there is a cyclone which relieves excess steam from the system. The cyclone pressure is regulated by a control loop using a transmitter, controller and exhaust valve. A constant flow of hot water enters the cyclone to maintain correct consistency. The pressure control loop on the secondary refiner pulp line is the same as on the primary refiners.

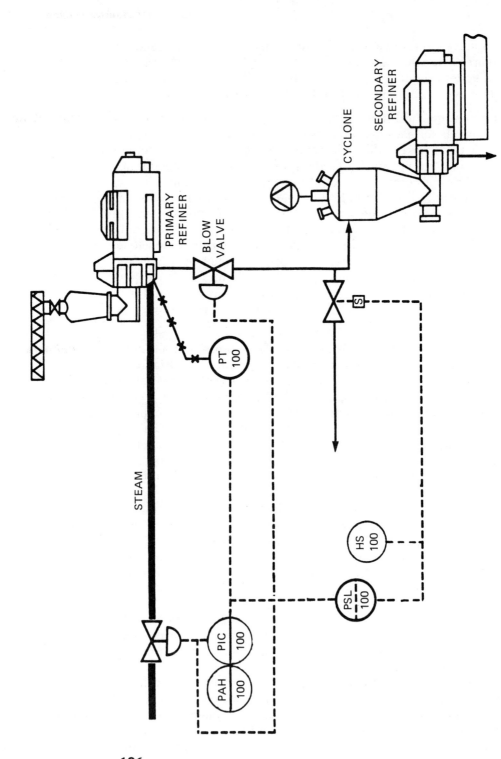

Figure 5–13. Refiner pressure control system.

The pulp leaving the secondary refiner flows into a pressurized cyclone on the output of which is a conical screw. The conical screw is used to maintain the upstream pressure in the refiners. The screw is of the same design as the one used before the cooker except the speed of rotation is constant. The cyclone removes excess steam from the pulp stock and uses a pressure control loop with an exhaust valve to maintain constant pressure. As in the previous cyclone, there is a constant flow of dilution water present to reduce pulp consistency.

Conical Screw

The conical screw feeds the pulp to the latency chest where a constant pulp supply is stored and held for around nine hours. Pulp consistency is verified manually or automatically to guarantee a high quality pulp. When measured manually, the pulp consistency is checked every two hours and the white water control set point adjusted to the desired value. Since the consistency is measured in the laboratory, the readings cannot be done more often. Once the system is running without interruption, consistency is only checked every four or five hours. When the system's consistency is verified automatically with a consistency control system, manual verification must still be performed from time to time.

INSTRUMENT LIST

	Tag No.	Description
Screw conveyor control system	LT 100	Pressurized screw conveyor level transmitter
	LIC 100	Level controller
	LALH 100	High and low level alarm
	LY 100	Level signal adder
	HC 100	Auto manual switch
	LZ 100	Screw speed control transducer
	LZ 101	Screw speed control transducer
	LALH 101	High and low level alarm
	LIC 101	Level controller
	LT 101	Cooker chip level transmitter
	ST 101	Screw speed transmitter
	ST 102	Screw speed transmitter
	SY 101	Screw speed adder
	SIC 101	Speed controller
	SIC 102	Speed controller
	SZ 101	Screw speed control
	SZ 102	Screw speed control
	FIC 200	Flow controller, remote set point with bias
	FIC 201	Flow controller, remote set point with bias
	FIC 202	Flow controller, remote set point with bias

Tag No.	Description
FT 200	Flow transmitter
FT 201	Flow transmitter
FT 202	Flow transmitter
FY 200	Current to pressure transducer
FY 201	Current to pressure transducer
FY 202	Current to pressure transducer
PT 100	Pressure transmitter
PIC 100	Pressure controller
PAH 100	Pressure high alarm
PSL 100	Pressure switch low
HS 100	Hand switch

(Refiner pressure control system)

PULP CONDITIONING PROCESS

Once the pulp has been refined, it is still not paper machine quality. For this reason the pulp must be treated to remove dirt, separate larger fibers from shorter ones, remove hard fibers and adjust pulp consistency (see Figure 5–14).

PULP CONDITIONING EQUIPMENT

Pulp Screens

Depending on the desired pulp quality and the system capacity, there could be any number of primary, secondary, and reject screens. Since the screens in each group are the same, we need only examine one group.

The latency chest supplies pulp to the retention chest which is then pumped to the screens. The primary screen serves to separate acceptable from nonacceptable pulp. The pulp is pumped at high pressure into the primary screen. Here it enters from the top into the middle of the spinning screen which is filled with small holes. The larger fibers cannot get through the holes, so they stay in the center where a flow of white water forces them to the bottom and out of the screen for further treatment. The fine pulp goes through the screen holes and is pumped to the primary cyclone cleaner retention chest.

The percentage of material rejected from the pulp screen is established by laboratory testing of the pulp before it enters the thickening drums. This percentage is controlled by the flow controllers on the reject lines. The rejected pulp is pumped from the secondary pulp screen retention chest to the secondary pulp screens. The accepted pulp from the secondary pulp screen flows into the primary pulp screen retention chest and is recycled through the primary screen.

Screw Press

The rejected material from the secondary pulp screen ends up in the nonrefined pulp reject chest. From here, it will be pumped to a screw press to increase consistency and then sent into a reject refiner. Once the rejected pulp has been refined again, it is pumped

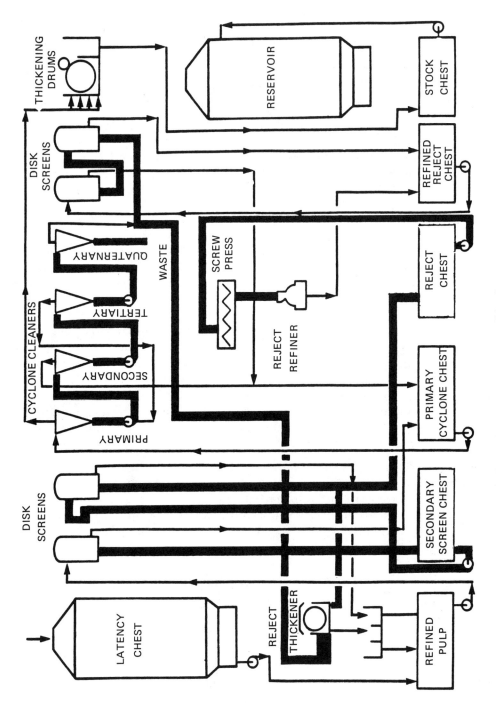

Figure 5–14. The pulp conditioning process.

199

into the refined reject pulp chest. From this chest the pulp is pumped into the primary reject screen. The accepted fibers flow to the primary cyclone cleaner retention chest from here, and the rejects go to the secondary reject screen. In the secondary reject screen the accepted fibers are recirculated to the primary reject screen and the rejects are sent back to the reject refiner.

Cyclone Cleaners

Cyclone cleaners are used to eliminate heavy fibers from pulp stock (see Figure 5–15). These fibers are not desirable because they reduce the quality of the paper. They are hard pin fibers and sometimes dark in color. In most systems there are a great number of cyclone cleaners and it is not surprising to see over 100 of these on a pulp line.

From the primary cyclone cleaner chest, the pulp is pumped into the primary cyclone cleaners. The main line splits up to accommodate all the cyclone cleaners. In the cyclone, the heavier fibers fall to the bottom into a trough from which they will be pumped to the secondary cyclone cleaners. The accepted fibers which exit through the top of the cyclone flow to the thickening drums.

From the secondary cyclone cleaners, the accepted fibers flow back to the primary cyclone cleaner retention chest. The rejects are pumped into the tertiary cyclone cleaners. From here the pulp either ends up back at the secondary cyclone cleaners or the rejects go to the quaternary cyclone cleaners. The rejects from the quaternary cyclone cleaners are thrown into the trash pile. The accepted pulp is pumped to a thickening drum and from there to the nonrefined pulp reject chest.

Thickening Drum Filters

At this stage the pulp has reached acceptable quality for use on the paper machine. However, one change that must still occur is an increase in the pulp consistency.

A drum revolves in a pulp bath. The center of drum is under vacuum pressure and as it turns pulp sticks to it. This causes water

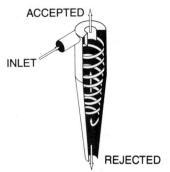

Figure 5–15. Cyclone cleaners.

to drain inside the drum. A rubber roller removes the pulp from the drum and a scraper removes pulp from the rubber roller. The pulp is then pumped to the stock chest. The residual water is re-circulated to the white water chest.

After the stock chest, the pulp is put through a bleaching process. At this point the pulp is ready to be transferred to the head box at the paper machine.

Bleaching Process

The bleaching stage alters the color of the pulp used to form the sheet of paper. In this stage a variety of chemicals can be added to whiten the pulp. Most of the chemicals arrive at the plant in bulk. This means that the plant must mix and prepare the final solutions, which involves mixing different chemicals together using conductivity, density and pH measurements as variables to control the final product.

The different chemicals that can be used in this process are, among others, calcium hypochlorite, sodium hypochlorite, lime, and chlorine dioxide, etc. Also a variety of additives may be put in at this point to change the final color or some other characteristic of the paper.

PULP CONDITIONING CONTROL SYSTEM

Primary Pulp Screens

As the pulp is pumped from the latency chest, a flow transmitter and indicating totalizer keep track of the quantity of pulp leaving the latency chest. Downstream from the flowmeter is a control valve positioned by a controller which regulates the level of the pulp screen retention chest. This loop also contains a high and low level alarm to warn the operator of these conditions.

From the retention chest, the pulp is pumped to the primary pulp screens. The flow of pulp stock through the screens involves control systems with remote set points (see Figure 5–16). Two flow transmitters are used on each screen. One is on the rejected line and the other is on the accepted line. The signal from the accept pulp transmitter serves as a remote set point to the flow controller on the reject pulp line. On the reject pulp flow controller there is a low flow alarm. The accept line flow controller receives its remote set point signal from a level controller at the primary cyclone cleaners retention chest.

If there is more than one primary pulp screen, the remote set point on the accept pulp line flow controllers is common. On both the input line and the accept line there are pressure indicators which monitor screen efficiency. On all the screens there is a flow transmitter, controller and valve on the white water input line.

On the retention chests are located level transmitters with high and low level alarms. These safety features warn the operator of any possible malfunctions of the control loops or process material supply.

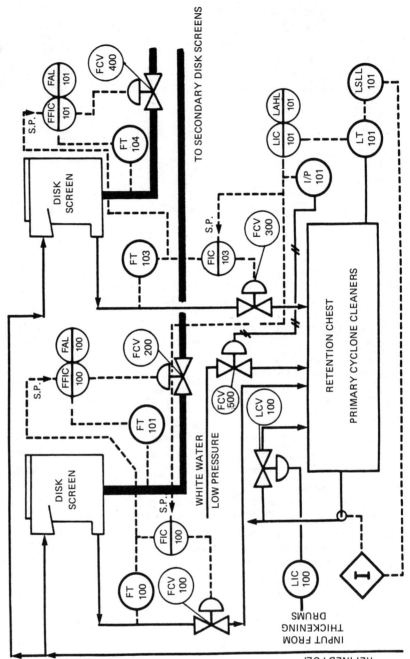

Figure 5–16. Level control for pulp disk screens.

202

On the cyclone cleaners themselves there are no control loops, only pressure indicators and high pressure switch logic solenoids to interrupt input flow. Pulp flow to the bank of cyclones is controlled by a level controller at the thickening drum filters (see Figure 5–17). The level controller signal opens and closes a valve on a bypass line that recirculates the pulp flow back into the retention tank from the main line. There are high and low alarms on the level of the thickening drum filters. Above the thickening drum filters there is a camera for observation of the drum filters' performance by the process operators in the control room.

This brings us to the stock chest. From here pulp is pumped into storage tanks where it is kept until needed at the paper machine.

Let us look at a calcium hypochlorite bleach process control system (see Figure 5–18). This process involves diluting lime with water to obtain a correct density and then adding to it vaporized chlorine. This results in calcium hypochlorite bleach production that will be used for bleaching pulp.

Lime arrives at the plant either by truck or train. From the hopper, lime is pumped into a slaker where it is crumbled by adding water with a manually set speed control pump. The pump is remotely switched by a level controller on the slaker that also switches a solenoid on the water dilution line. The slaker temperature is regulated by means of a temperature controller and a control valve on the steam line.

Next to the slaker there is a degritter that removes grit from the lime. From the degritter the diluted lime is pumped through a cyclone into a lime slurry tank. After the cyclone, the lime's density is measured and controlled by adding water to the degritter. The level in the lime slurry tank is maintained by a level controller and a valve on the tank inlet. From the lime slurry tank, the lime is pumped through another cyclone and into the dilute lime tank. At the cyclone input, density is measured and controlled by adjusting the water flow to the lime pump inlet. The level in the dilute tank is regulated by a level controller whose signal goes to a three-way valve that can reroute the lime flow back to the lime slurry tank.

From the dilute tank, the lime is pumped into a recycle tank. From here the lime is pumped to a reactor that mixes chlorine gas and lime to create the final bleach solution. The level in the recycling tank is maintained by a level controller and a valve on the inlet line from the dilute tank. Following the reactor stage, the bleach goes through a cyclone and from here the rejects either go to a degritter and back to the recycling tank or out to waste.

On the cyclone output there is an oxidation-reduction potential (O.R.P.) controller. It indicates chemical reaction completion. The

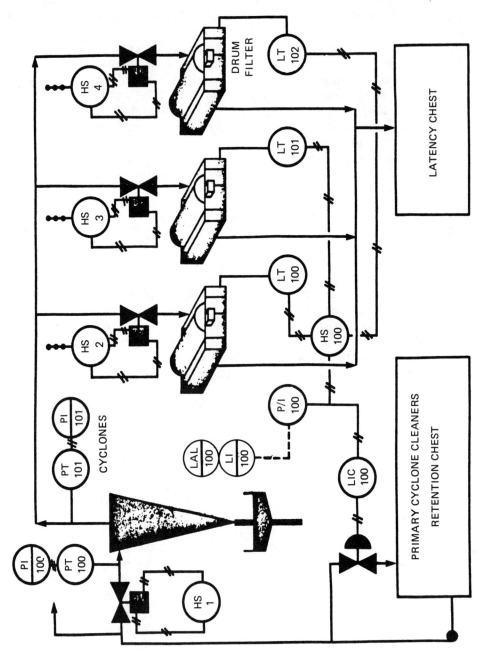

Figure 5–17. Cyclone cleaners and thickening drum filters control system.

204

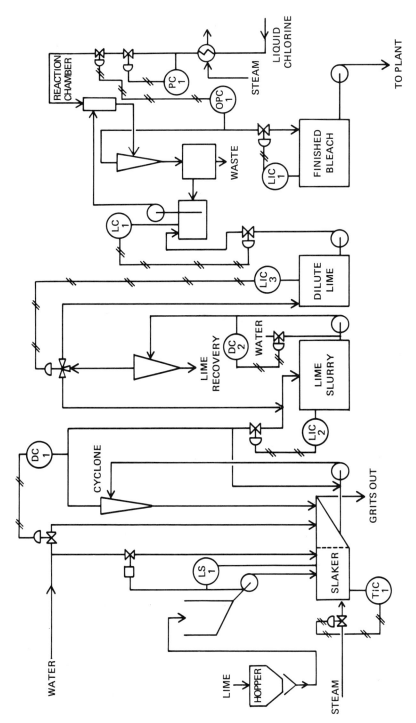

Figure 5-18. Instrumentation for a calcium hypochlorite process.

205

O.R.P. controller will either increase or decrease the flow of chlorine to the reactor in order to maintain reaction completion.

Since the chlorine is delivered in liquid form, steam heat is used to vaporize it. Upstream from the oxidation-reduction flow control valve, there is a pressure control loop which keeps the pressure of the chlorine gas constant.

The level in the finished bleach tank is set by a level controller and a control valve on the bleach inlet line. Once the bleach has reached the finished bleach tank, it is ready for bleaching pulp stock.

INSTRUMENT LIST

	Tag No.	Description
Level control for pulp and disk screens	FT 100(3)	Accepted pulp flow transmitter
	FIC 100(3)	Accepted pulp flow controller, remote set point
	FCV 100(300)	Accepted pulp flow control valve
	FT 101(4)	Rejected pulp flow transmitter
	FFIC 100(1)	Rejected pulp flow controller, remote set point
	FAL 100(1)	Low reject pulp flow alarm
	FCV 200(400)	Reject pulp flow control valve
	LT 101	Primary cyclone cleaners level transmitter
	LSLL 101	Critical low level switch
	LIC 101	Pulp level controller
	LAHL 101	Pulp level high and low alarm
	I/P 101	Current to pressure transducer
	FCV 500	White water flow control valve
	LIC 100	Thickening drum level controller
	LCV 100	Thickening drum flow control valve
Cyclone cleaners and thickening drum filters control system	PT 100(1)	Pressure transmitter
	PI 100(1)	Pressure indicators
	LT 100(1,2)	Thickening drum filters level transmitter
	HS 100	High selector
	LIC 100	Level controller
	P/I 100	Pressure to current transducer
	LI 100	Level indicator
	LAL 100	Low level alarm
Instrumentation for a calcium hypochlorite process	TIC 1	Slaker temperature controller
	LS 1	Slaker level switch
	DC 1	Density controller
	LIC 2	Lime slurry level controller

Tag No.	Description
DC 2	Density controller
LIC 3	Dilute lime level controller
LC 1	Level controller
LIC 1	Finished bleach level controller
ORC 1	Oxidation reduction controller
PC 1	Pressure controller

THE PAPER MACHINE

Once the pulp has been refined and conditioned, it is ready to be formed into a sheet of paper by the paper machine. The paper machine is divided into two sections: the wet end and the dry end. The wet end involves the head box reservoir, the sheet forming section, and the pinch roller press. The dry end is composed of drier rollers, the calender, and the roll cutting and wrapping section.

In certain cases, before reaching the head box, several types of pulp are blended to obtain different grades of paper. Depending on the quality of paper required one can mix thermomechanical pulp with chemically made pulp or pulp made of different wood chips shipped from other mills in order to attain the desired results.

THE WET END

The pulp is pumped into the head box and from here it is ejected through the head lip between the paper former's two screens. These carry the low concentration of pulp and form it into a paper sheet. The two screens squeeze the jet of pulp, removing some of the water so the pulp sheet can hold its shape.

At the three pinch roller press, the pulp is pulled through three felts which carry the pulp between a series of rollers. This causes even more water to be squeezed from the pulp and at the press exit the sheet holds itself together. From here the sheet goes to the drier section.

THE DRY END

The drier is where the pulp becomes a firm sheet of paper. From this stage, it will go through the calender which smoothes out the sheet. The drier constitutes one of the most expensive sections of the paper mill. As many as forty-five rollers heated by steam are used in five drier stages.

In the calender, the paper sheet receives it finish and thickness specified for production. The calender rollers are mounted vertically as opposed to horizontally as in the rollers in the drier stages. The distance between the calender rollers is adjusted to meet the required paper thickness. From the calender the finished sheet of paper is wound onto rolls which are ready for cutting and shipping to the customer.

The paper, now in rolls, is ready for shipping. It was wound at a precise moisture level, so the paper gives good results at the printing presses. This moisture level must be maintained until the customer receives the paper. An alternate method of shipping is in the form of paper sheets. This involves modification of the roll winding stage since the paper has to be cut into sheets instead of wound into rolls. This, however, is only done in mills that specialize in sheet paper. Many mills wind paper and send it to a second party to make sheets from the rolls.

HEAD BOX
The head box contains the pulp stock and its purpose is to inject a jet of pulp into the paper former. There are many variations in head box construction. We will look at one called the *injection* type (see Figure 5–19).

In this case, the head box is always full of pulp. The supply of pulp into the reservoir is brought through calculated inlets. This creates a continuous turbulence in the pulp flow so that the fiber consistency is uniform throughout the reservoir.

From the reservoir the pulp enters a perforated plate which leads it into the multitube section. This section, consisting of rows

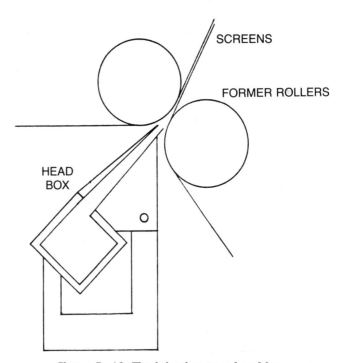

Figure 5–19. The injection type head box.

of tubes, serves to create a turbulence level high enough to keep the fibers dispersed and to stabilize the flow rate.

Once the pulp leaves the multitube section, it enters the parallel part of the head which converges into a lip from which is formed the pulp jet. This lip is adjusted so the pulp jet enters the screens at the correct angle to the rollers. The upper lip can be adjusted in such a way that allows control of both the angle and velocity of the pulp flow.

The entire head box (with the exception of the perforated plates) is made of 316 stainless steel for durability and is highly polished to prevent damage to the fibers during the process. The perforated plates are made of high density polyethylene.

On the head box, the control systems involve temperature, lip angle, opening (also known as *ruler control*) and the velocity of pulp flow (see Figure 5–20). Head box temperature should be adjusted so that the difference in temperature between the head and the reservoir does not exceed 30°F (16°C). Otherwise, thermal shock may occur and cause damage to the head box. The maximum temperature of the head box should not exceed 160°F (71°C). **Control System**

Control of the lip opening is accomplished by adjusting an air operated screw-jack. In order to have the correct angle of flow, there

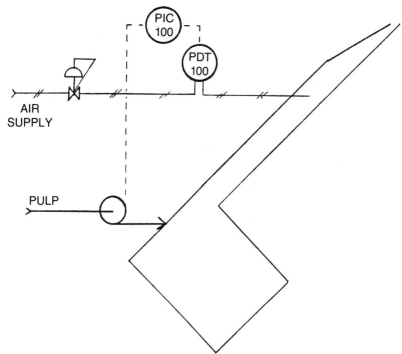

Figure 5–20. Head box pressure control.

is another screw-jack that is adjusted inside the lip. With a calculated distance between the two ruler measurements and the aid of a graph, the pressure in the head box can be found. Using a differential pressure transmitter, the velocity can be increased or decreased. The high pressure port of the transmitter is connected to a pressure detector in the head and the low pressure port to an air regulator. By changing the air regulator pressure, the transmitter output signal will vary and this allows a trimming adjustment at the paper machine by a local operator. The transmitter output is connected to a controller which varies the speed of the pulp pump on the inlet of the reservoir which changes the pulp flow velocity. An interlock system should be included which will not permit adjustment of the head box pressure if the former is not in operation since this can cause damage to the former screens.

SHEET FORMER At the head box outlet is the sheet former. This former consists of two screens which feed the pulp between rollers and create the pulp sheet. The jet of pulp lands between the two screens which squeeze the pulp jet to remove some of the water. The sheet begins to take form. The water removed is recirculated to the TMP process and reused in the system. The paper grade depends on the consistency of the pulp entering the sheet former, the speed of the former versus the velocity of the pulp flow, and the type of screens used on the former.

There are three essential rollers in the sheet former. These are the head and form rollers between which the initial flow of pulp enters and the couch roller which follows the form roller (see Figure 5–21). The couch roller gives the sheet its final consistency before it enters the drier stage. Throughout the former are located jets of water necessary to clean and moisten the felts which will keep the rollers and the machine in good working order.

THREE PINCH ROLLER PRESS From the sheet former, the sheet is fed to the three pinch roller press (see Figure 5–22). The sheet is transferred to the grabber felt from the lower felt of the former which leads the sheet between the grabber felt and the lower felt toward the first pinch rollers.

The first pinch rollers consist of an upper suction roller and a lower groove roller which removes water from the sheet. The sheet then goes to the next pinch rollers which consist of the previous upper suction roller and a lower granite roller. Here the sheet is no longer in contact with the first felt and follows the granite roller to the third pinch rollers. These consist of the granite roller and another groove roller for the final removal of white water. The third felt follows the sheet through the rollers, and from the third pinch rollers the sheet enters the drier stage of the paper machine.

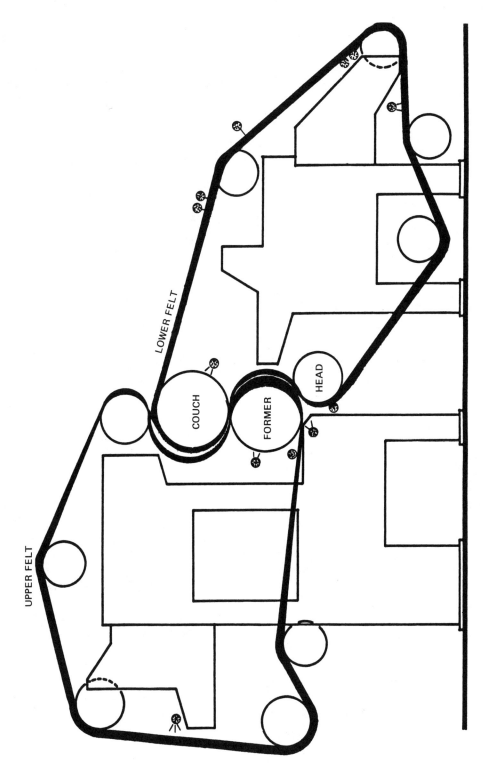

Figure 5–21. The three rollers of the former and the water jets.

UPPER FELT

LOWER FELT

COUCH

FORMER

HEAD

211

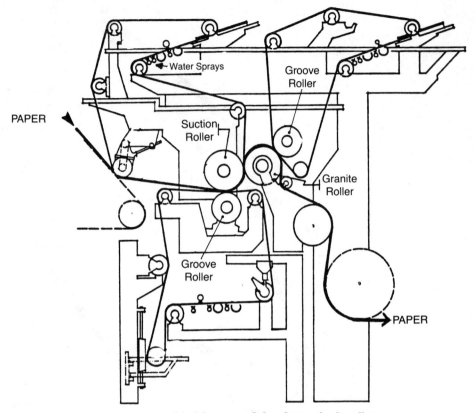

Figure 5–22. Diagram of the three pinch roller press.

DRIER   This stage consists of five sections of drier rollers. In the first two sections, spacing between the rollers is large. This wider spacing exposes a larger area of paper to air. In the other roller sections, spacing between the rollers is smaller and also more rollers are used than in the first and second sections.

The rollers used in the drier stage are made of metal and are hollow in order for steam to circulate and heat the rollers. The temperature of the rollers varies depending on their location with the first rollers being not as warm as the last rollers. This is done to prevent the paper from being heated too quickly. When that occurs, the outside edges will dry and wrinkle the paper, making it more difficult to dry in the later sections.

A paper guide system will feed the sheet between the rollers correctly whenever there is a break and the sheet must be refed through the system. This guide can either consist of wires that follow the path the paper sheet takes or scrapers which send the paper in the right direction.

Condensation is removed from the inside of the rollers by vacuum pumps. As the steam enters the rollers, it cools off against the rollers' inner walls and forms condensate. Not all steam entering the rollers condenses, so whatever steam exits the rollers with the condensate is redirected to the previous roller section, which operates at a lower temperature.

In the middle of the drier stage, a coating unit is sometimes added to change the finish on the paper. The coater can add gloss or any other desired finish. Just before the last roller, a paper cutting blade is used to trim the paper square so it can go through the calender easily.

The last roller in the drier stage is the moist roller. This roller is not heated like the drier rollers but is cooled and moisturized. This is done to insure an even moisture level across the sheet of paper.

Control System

The control system at the drier stage involves roller speed, steam pressure, condensation level, paper sheet moisture level and sheet break detection (see Figure 5–23). There are two steam lines going to the rollers. One is a high pressure line, the other a low pressure line. They are linked to the main distribution line that feeds the rollers of one section. The pressure in the main distribution line is measured and controlled by adjusting the valves on both the high and low pressure steam lines. These valves have a common control signal. At the high pressure valve, high pressure steam and the steam from the next section's condensate steam overflow are mixed together.

Between the steam pressure input to the rollers and the condensate pressure output from the rollers is a differential pressure controller. Its set point governs roller temperature since the larger the pressure differential the greater the flow of steam through the rollers and the higher the temperature. The output of the controller goes to two valves on the condensate tank. One valve allows excess steam out to the condensate heat exchanger and the other directs steam to the input of the previous roller section.

On the condensate tank a level controller regulates the valve on the output side of the vacuum pump. The level control, as opposed to constant flow control, prevents steam from going back to the boiler.

There are at least three paper break detectors in the drier and calender sections. One is at the input to the drier, one is located just before the last rolls on the drier and one is set after the calender. Speed control of the drier has always been a problem since each section of drier has an independent drive system. Digital control techniques solves this problem.

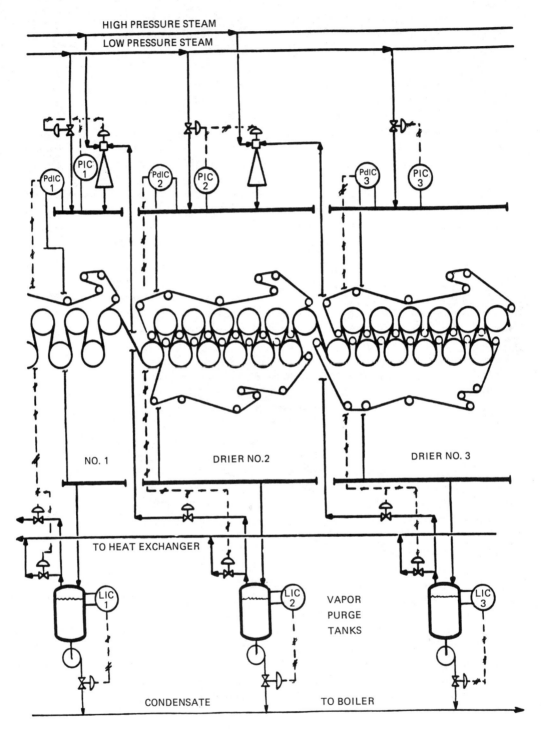

Figure 5–23. Steam control at the drier rollers.

To obtain an even thickness and a smooth finish for the paper, a calender is used after the drier section. The calender consists of a series of vertically mounted rollers ranging in number from five to twelve. Calender construction depends on the type of paper manufactured at the plant. Variations in calender design include the type of rollers used, the number of rollers, and the method of varying pressure across the contact surface. Rollers used on a calender can be solid, water heated or cooled, and their circumference can be varied by changing the oil pressure inside the rollers. Pressure on the rollers can be varied by changing the overall pressure on the top roller or on the roller axis.

CALENDER

Due to possible variation in paper tension and to keep a constant tension on the sheet paper entering the calender, a tension roller is positioned between the moist roller and the calender. This roller is motor driven and supported by air cushions. It acts as a shock absorber to prevent the paper from breaking during tension changes. Air pressure in the cushions can be changed to vary the tension of the paper.

In the case of a simple calender, control is usually manual. This control involves adjustment of the downward pressure exerted on the rollers. As shown in Figure 5–24, nip pressure is varied by adjusting the force on each end of the top calender roller independently. This is accomplished by loosening or tightening the loading arm which in turn exerts pressure on the top roller. This adjustment is not continuous but only has to be performed when the paper varies in thickness.

Calender Control

Control of a fully automated calender is more complex and specialized computer control systems are available for this application. Just after the paper sheet comes out of the calender, an on-line transmitter measures basis weight, caliper and humidity across the whole sheet. This information is then sent to the computer for analysis and any automatic appropriate corrective action necessary.

Using the above variables the control system will adjust the temperature of the water heated or cooled rollers and, in the case where oil filled rollers are used, vary the shape of the circumference of the rollers (see Figure 5–25). Not only can shape of the circumference be varied but also the pressure exerted on the roller ends can be changed. With this type of system the sheet can be of optimum quality. In some cases the rollers are not mounted in a perfect vertical line but are slightly offset. This helps in the calender control.

The rollers on the calender can be coated and made of different material. Also, there are often one or more rollers that are motor driven. Once the paper leaves the calender it is rolled, then unrolled, cut up to size, packaged and sent either to storage or to a customer.

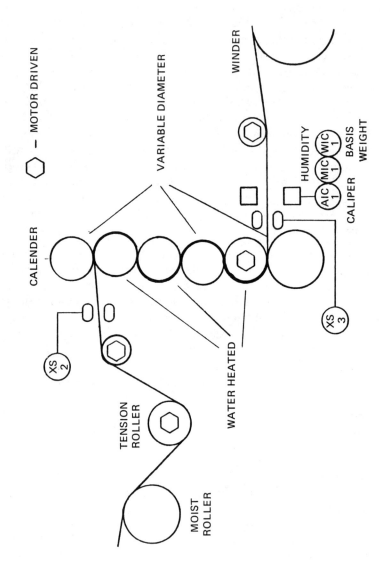

Figure 5–24. Calender rollers and control system.

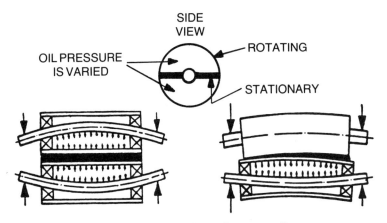

Figure 5–25. Variable circumference rollers.

INSTRUMENT LIST

	Tag No.	Description
Steam control at the drier rollers	PDIC 1(2,3)	Differential steam pressure controller
	PIC 1(2,3)	Inlet steam pressure controller
	LIC 1(2,3)	Vapor purge tank level controller
Calender rollers and control system	XS 2	Break detector switch
	XS 3	Break detector switch
	AIC 1	Caliper controller (sheet thickness)
	MIC 1	Paper humidity controller
	WIC 1	Basis weight controller

6 Textiles

DYEING OF TEXTILE YARN AND FABRIC

Dyeing of animal and vegetable fibers is a science more than several thousand years old. Before the advent of synthetic yarns and the color-conscious consumer, the operator of the dye tub controlled the bath temperature by hand. He used his trained sense of feel to judge the temperature. Results varied, certainly, but there was no consumer demand for a consistent product time after time. Once this demand occurred, however, greater accuracy was required and sensitive controllers were used. With these systems, the rate of rise (and cool) was still done by hand by slowly opening a steam valve, for instance. The hold time was determined by the operator watching a nearby wall clock.

To increase efficiency and improve quality in the textile dyeing industry, instrument manufacturers have developed temperature controllers capable of obtaining the required temperature profile with little attention from the operator.

THE DYE CYCLE

All dye cycles consist of loading the product in a bath into which chemicals and dyes have been or are to be added at some initial temperature. The bath temperature is increased at a certain rate for a set length of time. The product is then cooled, rinsed and withdrawn. Continued experimentation determined that most products could be dyed faster if the hold temperature was increased above the normal boil. To accomplish this it was necessary to increase the operating pressure inside the dye vessel. This required new equipment—equipment manufactured from stainless steel plates resistant to corrosion, with added strength and with the capability of operating under pressure.

DYE BECK

If the goods are circulated through the dye liquor by a rotating reel mounted above the liquor, as shown in Figure 6–1, then the vessel containing the liquor is termed a *dye beck*. It is either atmospheric or high-pressure. If the goods are in pieces rather than continuous,

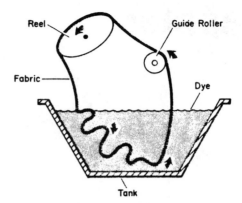

Figure 6–1. Dye beck.

the vessel is named a *piece dye beck*. The dye beck handles products from tubular cotton knits, 30 inches (75 cm) in diameter in rope form to 15 foot (4.5 m) wide carpet in either rope form or spread flat (termed *open width*).

PADDLE DYE MACHINE

Piece goods can also be dyed in a paddle machine. The goods are enclosed in mesh bags, placed in the dye bath (the container of which is a dye vat) and agitated by a paddle wheel until dyeing is complete. A paddle dye machine is shown in Figure 6–2.

BEAM DYE MACHINE

In a beam dye machine the material to be dyed is wound on a long (17 to 20 feet, or 5 to 6 m) hollow, perforated, stainless steel tube called the *beam*. The loaded beam is placed in the dye machine, horizontally or vertically depending on machine geometry. The beam is sealed on both ends by the action of an air cylinder, as shown in Figure 6–3.

When the beam is sealed, the dye liquor is circulated from the machine tank through the pump, and back to the inside of the beam. The liquor is forced through the perforations in the beam, through

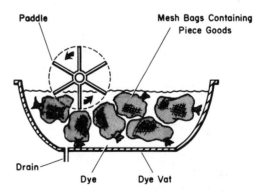

Figure 6–2. Paddle dye machine.

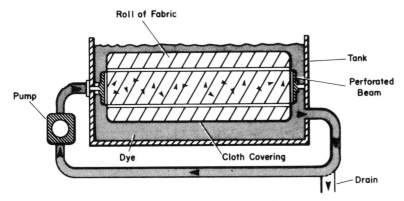

Figure 6–3. Beam dye machine.

the product on the beam and returned to the tank. For dyeing at elevated temperatures, the machine is closed, sealed and pressurized. The product wound on the beam is either strands of yarn or fabric of relatively open structure such as tricot.

Dyeing can also be efficiently done in a package dye machine, shown in Figure 6–4. Yarn is wound on a short, 6 inch (15 cm), perforated dye tube in approximately 1 pound (0.5 kg) packages. These packages are placed on hollow perforated spindles, 4 to 8 packages per spindle. The spindles are permanently vertically

PACKAGE DYE MACHINE

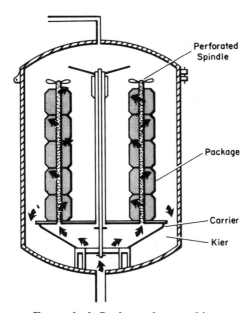

Figure 6–4. Package dye machine.

mounted on a circular manifold. This assembly is known as a *carrier*. The carrier with the packages to be dyed is placed upright in a vessel termed the *kier* and a pressure-tight lid is clamped on. The kier is capable of handling temperatures in excess of 265°F (130°C).

DYE JIG

Fabric can also be dyed on jigs or padders, shown in Figure 6–5. On a dye jig the fabric is run from one large roller through the dye and on to the second large roller after which the direction is reversed. In this manner the fabric can be run back and forth through the bath as is necessary to achieve the required shade. The tightness of wind on the roller tends to squeeze the dye through the fabric. Both open and closed jigs are used.

PADDER

The padder shown in Figure 6–6 consists of a shallow trough containing the dye liquor and two squeeze rollers mounted above it. The fabric is passed through the trough and between the squeeze rollers where the excess liquor is pressed out and the remainder penetrates the cloth. To obtain the required shade of color the cloth can be run through several padders arranged in a row.

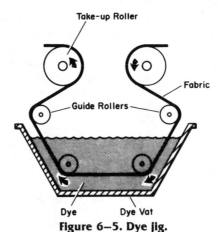

Figure 6–5. Dye jig.

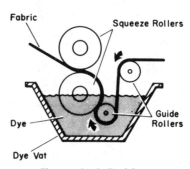

Figure 6–6. Padder.

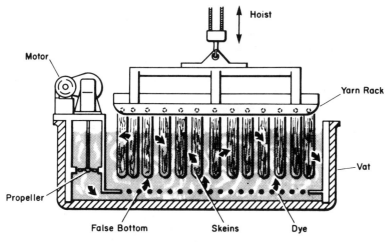

Figure 6–7. Skein dye machine.

Another type of yarn dye machine, shown in Figure 6–7, is a skein or hank dyeing machine, particularly used for woolen yarns. A skein or hank is a definite length of yarn wound in a coil. These coils are suspended by narrow rods on a yarn carrier and lowered into a dye tub. Circulation of dye liquor, in the vertical direction through the hanging coil of yarn, is accomplished with a motor driven propeller. The flow is changed by reversing the rotation of the propeller.

SKEIN DYE MACHINE

With continued experimentation, it was learned that if fabric could be freely floated in highly turbulent dye liquor, the dye cycle could be shortened due to increased penetration. This is the operation principle of a jet dyeing machine, shown in Figure 6–8. The cloth

JET DYE MACHINE

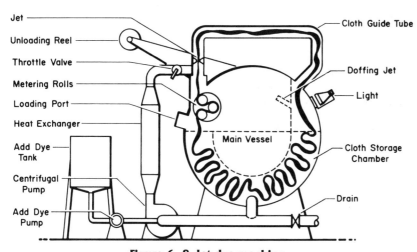

Figure 6–8. Jet dye machine.

is passed in rope form vertically upward through a circumferential converging jet of dye liquor which emerges from a ring-shaped slot completely surrounding the fabric. The fluid simultaneously penetrates and propels the fabric through the cloth guide tube. As the cloth is surrounded by a continuous ring of fluid which is moving faster than the cloth, the cloth is being floated through the tube with no cloth-to-metal contact. The machine is capable of being sealed so that dyeing can take place at elevated temperatures.

DRUG ADDITION

If the drugs (chemicals and dyes) are manually dumped directly into the dye tub, little is required in the form of equipment and communication, outside of mixing storage areas and portable tanks. If the drugs are to be added directly from tanks located in a central drug room, more sophistication is required. Some form of direct communication is necessary. It could be verbal via an intercom and/or lights and push buttons for each tank. With additional control sophistication, the tanks could be automatically dumped and rinsed. If desired, the tanks could also be filled and heated automatically, in preparation for the mixing of the drugs.

DYE LIQUOR TEMPERATURE CONTROL SYSTEM

The main process variable is the temperature of the dye bath. The time-temperature profile is shown in Figure 6–9. The mode of heat transfer in a dye machine is either by steam coils inside the tub or an external tube and shell heat exchanger.

The temperature control system has to be capable of transferring heat at a maximum required rate on the initial rise, without sacrificing good control during the hold period. The required heat input during the hold period is a fraction of that required on maximum temperature rise as there is no load change during the hold time and the heat loss is small. The above requirements can be accomplished by the use of two parallel valves—a larger one for temperature rise and a smaller one for use during the hold time.

DYE VESSEL PRESSURE CONTROL SYSTEM

When dyeing in a pressurized vessel, the head space is pressurized to keep the liquor from boiling as the bath temperature increases. Usually a pressure regulator is preset for the required pressure and a solenoid valve is energized to either vent or pressurize the vessel.

SEQUENTIAL

As in any batch process, the material has to be loaded, the process has to be completed and the material unloaded. The process sequence is such that the tub is filled to a set level with water, chemicals and dyes are added to the water and the bath temperature raised as described above. Depending on the equipment involved, there can be a pump and/or reel motors being turned on and off at the proper time.

When the material has been dyed to the required shade, the bath is dropped and the material rinsed. Again, depending on the

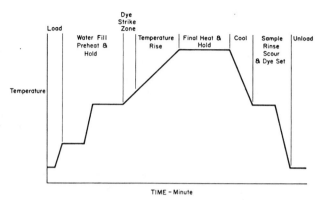

Figure 6–9. Typical time-temperature profile.

equipment, material type and configuration, the rinse cycles are not identical. Water can enter at the top of the tub and the depleted bath exits through the bottom drain valve until the bath is clear. Water can enter the bottom and overflow from the top. Variations of the above two methods are also used. The material handling is usually a manual operation while the remaining sequences can be accomplished manually, automatically or somewhere in between, depending on the specified sequencing controls.

BOIL CONTROL SYSTEM FOR ATMOSPHERIC DYE BECKS

In the textile industry, when fabric or yarn is batch dyed, the liquor temperature in the tub is controlled at a profile similar to that shown in Figure 6–10. When the dyeing equipment is capable of being pressurized, the boil hold temperature shown is well above the atmospheric boiling temperature. However, if the tub is open to the atmosphere, the hold temperature will be around the boiling point, or approximately 212°F (100°C).

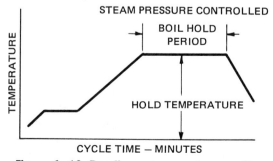

Figure 6–10. Dye liquor temperature profile.

In the carpet industry in particular, it is necessary during the hold period to have the dye liquor at a rolling boil to insure better dye dispersion. But with normal temperature control, a consistent rolling boil cannot be obtained from one day to the next due to barometric pressure changes. For example, if the barometric pressure changed from 29.5 to 30.5 inches of mercury (99.89 to 103.27 mbar), the boiling point of water would change from 211° to 212.7°F (99.5° to 100.4°C). If the hold temperature was set at the boiling point corresponding to the maximum barometric pressure (e.g., 213.6°F at 31 inches or mercury, or 100.9°C at 104.97 mbar), the rate of boil would be too great when the barometric pressure decreased. If a lower hold temperature was used, the rate of boil would be too low when the barometric pressure was high. One solution to the problem is to compensate the dye beck temperature measurement for barometric pressure changes. This system requires a high sensitivity absolute pressure measurement and an analog computation.

An alternate solution is to control the steam pressure, instead of dye liquor temperature, in the sparge pipe at a pressure set point which maintains the boil roll at the required level. Since a gauge pressure measurement is referenced to atmospheric pressure, compensation is inherent to the system.

CONTROL SYSTEM

The suggested control system is shown in Figure 6–11. During most of the cycle, the temperature of the dye liquor is controlled according to the profile programmed into the sequencer, KC 1. If, at the time the sequencer reaches a hold, the liquor temperature is close to the boiling temperature (above 208° to 210°F, or 98° to 99°C), the system will automatically transfer from temperature control to steam pressure control. The temperature alarm, TSH 1, senses the near-boiling temperature and energizes the boil set point positioner, HC 2, to convert the system to steam pressure control. At the end of the hold period, the system returns to its normal temperature control mode.

On a dye with front and rear steam sparge pipes, the temperature distribution remains nearly uniform as the steam valve opens wider during a rate of rise. This is due to steam entering from the two sparge pipes and creating a good mixing action. Some dye becks have only one sparge pipe, usually mounted along the front. As the steam valve is opened, the front of the bath is at a slightly higher temperature than the rear due to the action of only one sparge pipe. The temperature measurement is usually located in the front of the beck. Hence, it is indicating the higher front temperature at the start of the hold period as the system transfers to pressure control. As the hold period progresses, the bath temperature stabilizes due to continued mixing action. If the steam pressure in the sparge pipe remained constant, the boil rate would become excessive, since less

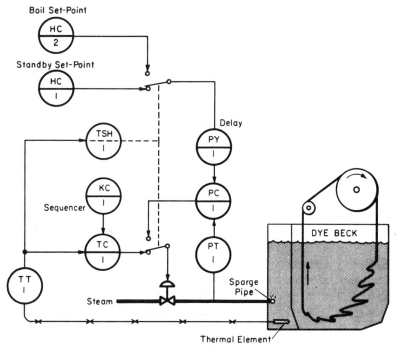

Figure 6–11. Atmospheric dye beck control system.

steam pressure is required after the bath temperature levels off. Therefore, to sustain a uniform boil rate during the entire hold period, it is necessary to decrease the steam pressure set point at the same rate that the bath temperature is stabilizing. As seen in Figure 6–11, the delay relay, PY 1, allows the pressure set point to decrease at an adjustable, preset rate nearly approximating the temperature lag in the stabilization of the bath.

The pressure controller is equipped with proportional response only, as process load changes are small and offset can be tolerated. The controller should be synchronized or manually reset after the system is in pressure control and the required rolling boil is obtained. Where the beck has one sparge pipe, the time lag in the delay relay, PY 1, is adjusted to obtain the required boil rate over the entire hold period. In a beck with two sparge pipes, the adjusted time lag is at or near zero.

Application Data

INSTRUMENT LIST

	Tag No.	Description
	KC 1	Bath sequence programmer
	TC 1	Liquor temperature controller
	TT 1	Liquor temperature transmitter
	TCV 1	Liquor temperature control valve
Pressure/temperature	TSH 1	Pressure switch/alarm unit
transmitter mode of	XS 1	Solenoid valve/relay
control system	XS 2	Solenoid valve/body
	HC 1	Standby set-point signal source
	PT 1	Steam pressure transmitter
	PC 1	Steam pressure transmitter
	PY 1	Steam pressure controller, set-point delay
		Electro-pneumatic transducer (for use with valve on electronic systems)

HIGH SPEED WARP SLASHER

Warp slashing in the textile industry is the process of conditioning the warp for maximum efficiency in the weave room. High speed looms subject the warp to chafing and rubbing due to the operation of the heddles, harnesses, and the shuttle itself. By the correct application of a size solution followed by controlled drying prior to weaving, the warp's physical properties are changed such that:

- Its strength, stiffness, and elasticity are increased due to the binding together of adjacent fibers, thus reducing breakage and slippage between fibers.

- Its fibers are forced to lay down the warp, making it more compact and smoother to reduce friction in the weave.

- Its external surface is covered with a comparatively hard size to increase its resistance to mechanical abrasion and to provide lubrication.

- Its ability to retain a static charge is reduced (depending on the type of size used).

The moisture content of the warp as it leaves the slasher is generally of great importance. By controlling the profile of temperatures on the slasher cylinders, a constant warp moisture is maintained.

Since a slasher may operate at speeds approaching 200 yards (180 m) per minute and supply warps for as many as 500 to 1000 looms, automatic controls to maintain consistency in the slashing process for highest efficiency in the weave room are essential.

Figure 6–12 shows a typical temperature control system for a single slasher cylinder, although any number of cylinders may be connected in parallel with common steam, condensate bypass, and condensate return lines. The common steam header feeding the cylinders is pressure regulated to eliminate upsets in can (cylinder) temperature control due to fluctuations in the plant steam source.

The control valve, TCV 1, modulates the quantity of steam into each can's rotary joint to heat the cylinder's surface. Condensate is withdrawn via a siphon connected to the rotary joint into a trap and condensate return common to all dryer can systems. The measurement of can temperature is fed to reverse acting controller, TIC 1, which modulates the air-to-open valve, TCV 1.

Where cylinders are operated near atmospheric pressure, an air purge is applied to the cylinder to eliminate large condensate accumulation in the can and to speed up the dynamic behavior of the control system. An air header is set near 5 psi (35 kPa). From the header to each cylinder journal there is an air connection with a

CONTROL SYSTEM

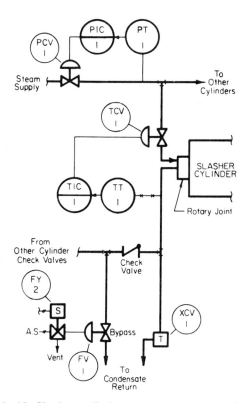

Figure 6–12. Slasher cylinder temperature control system.

ball check valve. When the pressure falls below 5 psi in the cylinder, air enters to create sufficient pressure to remove the condensate. When cylinder pressure is above 5 psi, the check valve prevents steam from backing into the air line.

Application Data

During startup, it is desirable to bring the cylinder temperature to its operating point as quickly as possible. As shown in Figure 6–12, condensate lines from the cylinders, with a swing check valve in each line, are connected to a common header located ahead of the traps. A diaphragm control valve located in the condensate header is operated by a pilot valve. This in turn, is regulated by the output of one of the cylinder temperature controllers. When the cylinder temperature is below the set-point temperature, the controller's output becomes great enough to energize the pilot and bypass rather than go through traps. This increases the steam throughput and causes the can to approach the control temperature in a shorter time. Near the control point the pilot valve closes the bypass valve and the condensate is removed through the traps. Even after startup, the control valve is automatically operated to vent the cylinders in case the temperature drops to a specified amount below the set point.

INSTRUMENT LIST

Tag No.	Description
PT 1	Steam supply pressure transmitter
PIC 1	Steam supply pressure controller
PVC 1	Steam supply pressure control valve
TT 1	Can temperature transmitter
TIC 1	Can temperature controller
TCV 1	Can temperature control valve
FV 1	Trap bypass valve
XCV 2	Bypass valve pilot valve
	Electro-pneumatic transducer (For use with valve on electronic systems—two required)

7 Miscellaneous Processes

GAS SCRUBBING TOWER

Boiler flue gases often contain harmful chemicals and particles which end up, by gravity, back on the earth's surface either dissolved or mixed with rain. One of the more harmful chemicals which forms acid rain is sulfur dioxide (SO_2). This gas is produced from the combustion of fuels such as oil or coal containing sulfur. One way to prevent these gases from entering the atmosphere is to practice in the plant what happens outside. This means to combine SO_2 gas with a water spray to form a collectable acid rain. This process is accomplished with a gas scrubbing tower.

GAS AND PARTICULATE SCRUBBING SECTION

In the boiler stack, water sprays are installed which produce showers of water similar to nature's rain. The shower of water absorbs SO_2, other gases and any particulates which fall to the bottom of the stack where they are collected and neutralized with a caustic solution.

THE HEAT RECOVERY SECTION

The scrubber also contains a two-stage energy recovery system in the upper part of the stack. Water is sprayed into the stack to collect heat from the hot gases leaving through the stack. First, water is pumped through water sprays in the highest part of the stack and collected in a tray at the next lower level. From here, to increase the efficiency of the heat recovery process, the water collected is once more sprayed just below the first collector. This water is again recovered and from here pumped into the plant as hot water. This water is relatively clean since the pollutants have been removed lower in the stack by the gas and particulate scrubbing section.

231

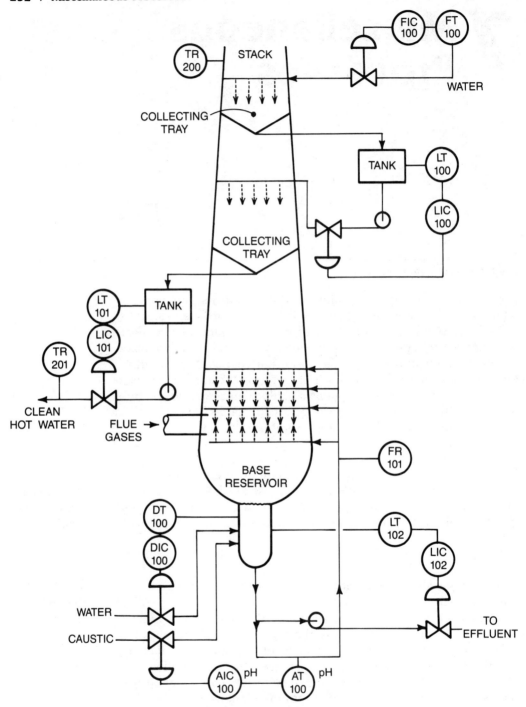

Figure 7–1. The gas scrubbing tower control system.

In the heat recovery process, the uppermost set of showers releases water as a fine mist which is heated by the flue gases. As the water falls it is caught in a collecting tray (see Figure 7–1). From this tray, water flows into a level tank where level transmitter LT 100 monitors the water level. Controller LIC 100 receives the transmitter's signal and controls a valve on the output of the collecting tray holding tank. These showers spray the water below the first collecting tray and more heat energy is recovered from the flue gases. Once again the water is collected in a tray and flows into a hot water holding tank. Level transmitter LT 101 monitors the level in the tank and transmits a signal to controller LIC 101 which controls a valve on the tank output. This water is then used in the plant for heating or other purposes.

Below the heat recovery section is the gas scrubbing section. Density is controlled by the density controller DIC 100 which obtains a density measurement from the base reservoir and controls water flow addition. Fresh water enters the base of the collecting reservoir and flows to a circulating pump which feeds both the discharge line and the water sprays. In some cases there is a flow recorder on the water line leading to the water sprays. The discharge line flow is controlled by level controller LIC 102, whose process signal is obtained from LT 102 at the base reservoir. At the circulation pump there is a bypass which supplies process fluid to the pH analyzer AT 100 which transmits a pH signal to pHC 100, the pH controller. The controller in turn adjusts the flow of caustic soda entering the base reservoir and so neutralization of the acidic water takes place.

CONTROL SYSTEM

INSTRUMENT LIST

Tag No.	Description
FT 100	Heat recovery water flow transmitter
FIC 100	Flow controller
TR 200	Stack temperature recorder
LT 100	Collecting tray holding tank level transmitter
LIC 100	Level controller
LT 101	Clean hot water holding tank level transmitter
LIC 101	Level controller
TR 201	Clean hot water temperature recorder
FR 101	Water spray flow recorder
LT 102	Base reservoir water level transmitter
LIC 102	Level controller
DT 100	Base reservoir water density transmitter
DIC 100	Density controller
AT 100	Base reservoir pH analyzer transmitter
AIC 100	pH controller

CONTROL SYSTEM FOR FILM PROCESSING

The film finishing operation shown in Figure 7–2 is typical of processes in which the raw material is transported through a series of tanks to obtain the desired finished product. The material is prepared, treated, set or prehardened, rinsed, and then dried. Each processing phase may consist of more than a single tank. For example, the prepare phase may consist of wash, bleach, and rinse tanks. Some film processing systems require nine different solution tanks. The major variables which cause differences in the finished product are the movement speed of the material and the quality of the processing tanks' solutions.

The speed of raw material through the various solution tanks determines how long the material will be in the tank. If the speed is constant the solution replenish rates can be fixed. If the speed is variable for different raw materials, the solution replenish rates can be set from the speed measurement. At slow speeds, the tank solution may be weaker for the same reaction on the raw material than at high speeds. Replenishment is necessary to replace expended and transported chemicals.

CONTROL SYSTEM

System Description

Two replenishment loops are shown in Figure 7–3. Two solution flow loops are illustrated with a single motor M driving the two solution pumps. Each solution is pumped from storage to the various processors at a constant pressure. The pressure is maintained by a conventional pressure control loop. Each processor has a recirculating pump for the solution with a conventional temperature control loop when necessary.

The solution line is connected to a three-way manual valve, HCV 1, HCV 3, at each processor. These valves are used for selecting quick fill, replenishment, or no flow. The replenishment lines contain a flow indicator, FI 1, FI 2, and a needle valve, HCV

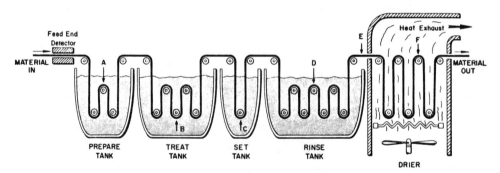

Figure 7–2. Film finishing.

2, HCV 4, for adjusting the replenishment flow rate. The sequencer KJC 1 controls the replenishment pump motor M and the solenoids, FY 1, FY 2, in the individual lines. With the replenishment rate adjusted to replace the expended and transported chemicals, the chemical concentration of the tank solution may be maintained at a consistent value, if the replenishment flow is started when the head end of the material is in the middle of the tank and stopped when the tail end of the material is in the middle of the tank.

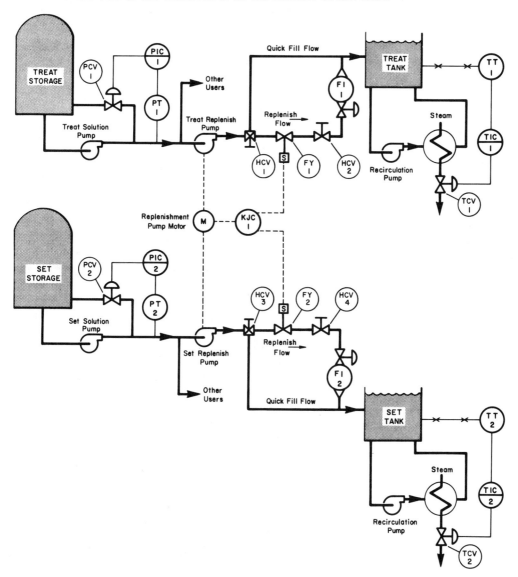

Figure 7–3. Typical replenish loop.

A material detector at the feed end of the processor is required. This provides the sequencer KJC 1, shown in Figure 7–3, with material head end and tail end information to start and stop the *on* and *off* sequencing cycles. The only other input to the sequencer is material speed. Since the material path is fixed for the processor, the time to the middle of any tank is inversely proportional to the material speed. The *on* sequencing cycle is begun when the material detector indicates the head end of the material. The *off* sequencing cycle is begun when the material detector indicates the tail end of the material. Both the *on* and *off* sequencing cycles may operate simultaneously if the length of material in the processor is shorter than the material path. The time interval counter for each sequenced device is set for the amount of time from the material detector to the center of the tank at the normal operating speed.

Contact outputs are provided in the sequencer to operate solenoids in the replenishment and rinse lines for pressurized solutions. Separate contacts are provided for starting pumps for nonpressurized solutions.

An interlock circuit is provided when two solution pumps are driven by one motor, as shown in Figure 7–3. When the sequencer turns on the motor M, solution is pumped from treat storage to the treat processing tank. The set replenish pump is pumping against a closed valve. To prevent excessive frothing and/or cavitation, solenoid valve FY 2 is opened after a fixed time interval. The interlock circuits operate when material is being processed at speeds slower than normal.

System Operation

Initially, the sequencer is set with the time for each output function. The processor tanks are prefilled to the proper levels and the recirculation pumps are turned on. Replenishment flow rates have been adjusted to the proper values. The processor transport system is turned on to the desired speed.

As shown in Figure 7–2, the material is placed in the feed end of the processor which causes the head end of the material to operate the material detector, initiating the *on* cycle. The sequencer opens the prepare replenishment solution line (not shown in Figure 7–2) after a time interval based on the speed when the material head end is at point A in the prepare tank. When the head end of the material reaches point B in the treat tank, the motor M shown in Figure 7–3 is energized and the solenoid FY 1 in the treat replenishment line is opened. The set solution is pumped but is not flowing until the head end reaches point C in the set tank. Then the solenoid FY 2 in the set replenishment line is energized. At point D, the rinse solution solenoid (not shown) is energized. At point E the heater and fans are energized for drying the material. Point F has no effect during the *on* cycle. The solenoids, pumps, heater and fans remain energized as long as material is running through the processor.

When the tail end of the material is detected, the *off* cycle is initiated. Replenishment solenoids are de-energized at points A, B, C and D. The motor M shown in Figure 7–3 is de-energized when point C is reached. Point E has no effect during the *off* cycle. When the tail end reaches point F, the heaters are turned off, but the fans are left running for a fixed interval to expend the latent heat in the dryer. Entering another piece of material at the feed end of the processor causes the above sequence to be repeated.

For applications in which the material is always longer than the material path in the processor and the processor is operated at only one speed, a single digital set programmer with *on* and *off* cycles will control the sequence cycles.

Application Data

For applications in which the material may be shorter than the processor material path and the processor is operated at only one speed, two digital set programmers are required with one programmed for the *on* cycle and one programmed for the *off* cycle since both may be operating simultaneously.

For applications in which the speed is a variable, the sequencer programmer is required to control the sequence cycle. The processor control panel may contain the controls for automatic sequencing, manual sequencing, and sequencing simulation. Status lights can also be mounted on the control panel to indicate solenoid and pump operation for monitoring the sequencing system.

INTRODUCTION TO REACTOR CONTROL

Chemical reactors are widely used throughout the process industries for the manufacture of synthetic rubber, polyvinyl chloride (PVC), polystyrene, epoxy resin, dye-stuffs, pharmaceuticals and food products. Chemical reactors can be classified as batch type or continuous type, and vary in capacity from 5 gallons (19 l) to greater than 4000 gallons (15 000 l). Their construction material is determined by the corrosive properties of the process reactants and products.

Close control of vital operating parameters is essential to insure the safety of operating personnel and plant property, to maintain consistent quality, and to minimize the cost of labor per product pound. Temperature, pressure, reactant flow rate and degree of reaction completion have been used as the primary process variables for control of the batch reactor. Temperature is often the most important variable and must be carefully regulated throughout the reaction cycle. This section is an introduction to the principles of chemical reaction, the construction of the batch reactor, principles of heat transfer and temperature control problems in a batch reactor.

Other sections deal with specific reactor control systems. These subjects include:

- Control of batch type chemical reactors—pressurized water circulation—control systems from a simple single loop with split range valves to a differential temperature control system for glass lined vessels.

- Control of batch type chemical reactors—external heat exchanger—cascade control system using pressurized oil heat transfer medium and external heat exchanger for cooling.

- Control of batch type chemical reactors—reactor startup—procedure for starting up and tuning batch chemical reactor with cascade control system.

- Control of batch type chemical reactors—time based program control—cascade control system with programmed master temperature.

CHEMICAL REACTION

The chemical reaction may be thought of as the bringing together of two or more raw materials or reactants under the influence of temperature and pressure in order to form one or more products. These products then possess different properties or characteristics than the reactants.

In general,

$$aB + bB = cC \pm H$$

where (a) moles of reactant (A) are added to (b) moles of reactant (B) to form (c) moles of product (C) plus or minus a quantity of heat (H). A *mole* is defined as the molecular weight of any substance, expressed in pounds or grams.

Reactions which generate heat while forming products are *exothermic* reactions. Reactions which require heat to form products are *endothermic* reactions. The quantity of heat generated or required in a reaction is determined from the reaction equation using the formation heat of the reactants and products. The rate at which heat is generated or absorbed is dependent upon reaction kinetics and is a function of the chemical nature of the reactants, the temperature, the reactant surface area, catalysts and the reactant concentration.

Chemical reactions can be categorized by the method in which they are carried out, such as:

- *Batch Reactions*—measured quantities of reactants are brought together in a vessel or reactor and allowed to react to form products. On completion of the reaction, the products are discharged from the reactor and the cycle is repeated.

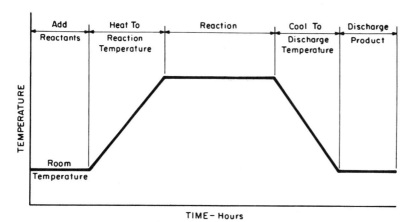

Figure 7–4. The batch reaction.

- *Continuous Reactions*—measured quantities of reactants are continuously passed through a reaction vessel at a controlled rate and products are continuously withdrawn.

Batch Reaction Cycle

The batch reaction cycle is conducted according to a definite time-temperature cycle. Figure 7–4 illustrates a typical cycle of many industrial processes.

Reactants are added to the reactor at room temperature. The reaction mass is then heated to reaction temperature and the reaction is carried out to form products. The batch is cooled and the reactor emptied. Total batch time in reactors used in the manufacture of PVC or polystyrene, for example, is approximately fourteen to eighteen hours.

BATCH REACTOR

Batch reactions are usually carried out in a closed vessel or reactor. Often, the product is discharged through a nozzle located in the reactor bottom. Figure 7–5 shows a commonly used type of jacketed batch reactor.

An agitator is used inside the reaction vessel to insure uniform mixing and to aid in the transfer of heat into or away from the batch as required. The agitator shaft passes through a nozzle at the top of the reactor. This nozzle has a stuffing box or mechanical seal to insure a tight seal when the reactor operates under pressure. The agitator, single speed, two speed, or variable speed, is driven by an electric or a hydraulic motor located above the reactor.

It is often necessary to add or remove heat from the reaction mass during various steps in the reaction cycle. To accomplish this, reactors have jackets or pipe coils through which a heating or cooling medium, such as steam, water or heat-transfer oil, is circulated. Figure 7–5 shows that this jacket is an outer shell which surrounds the main body of the reactor to form an annular space between the

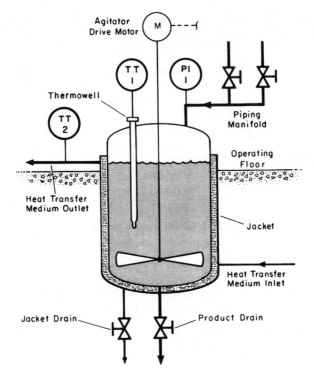

Figure 7–5. Jacketed batch reactor.

inside jacket wall and the outside reactor wall. The bottom of the jacket is formed around the bottom nozzle of the reactor. The jacket has flange connections that permit the connection of external piping and a drain can be provided at a low point so that the jacket can be emptied of all fluid. In practice, the sides and bottom of the reactor are covered with insulation.

In place of a full jacket, some companies use a full- or half-pipe coil to carry the heat transfer medium. In this type of installation, a pipe is wrapped around the outside of the reactor and welded to the reactor wall. Half-pipe coil jackets are formed in a similar manner to full-pipe coil jackets except that a pipe split lengthwise is used. Connections are provided in the jacket piping or the installation of thermo-wells and pressure gauges.

The top or head of the typical batch reactor is provided with flanged nozzles which are used for connecting the process piping, thermo-well and agitator. A manhole, to gain access to the reactor, is also provided.

Heat Transfer A review of some of the basic heat transfer principles is helpful in understanding specific control applications which are presented in other sections. Heat always flows from a higher temperature to a lower temperature. If an exothermic reaction takes place in a reactor and it is required to maintain the reaction mass at a constant temperature, the cooling medium in the jacket must be at a temperature lower than that of the reaction mass. The reaction mass can never be cooled below the minimum available temperature of the cooling medium. Conversely, the reaction mass cannot be heated higher than the highest available temperature of the heating medium.

The rate of heat removal or addition to the reaction mass is a function of the reaction vessel design, the difference in temperature between the batch and jacket, and the overall heat transfer coefficient. This is expressed mathematically by the equation:

$$Q = UA\Delta T$$

where
Q = rate of heat transfer,
U = overall heat transfer coefficient,
A = heat transfer area of jacket,
ΔT = difference in temperature between reaction mass and jacket.

The overall heat transfer coefficient is affected by the material used to construct the reactor, the physical properties of the reactants, the heat transfer medium, the agitation of the batch, and the rate of circulation of heating or cooling fluid through the jacket. Values of U for estimating the rate of heat transfer can be obtained from various reactor manufacturers or from publications such as the *Chemical Engineer's Handbook*.

To estimate the heat removed or added to the fluid in the jacket, use the equation

$$Q = C_p M \Delta T$$

where
Q = heat given off or absorbed,
C_p = specific heat of heat transfer medium,
M = flow rate through jacket,
ΔT = change in temperature of heat transfer medium.

To maintain a constant temperature during an exothermic action, the amount of heat generated by the batch must be transferred through the reactor wall and into the jacket. No control system can compensate for inadequate heat transfer surface or poor heat transfer characteristics.

CONTROL SYSTEM

The kinetics of the chemical reaction, the cyclic operation of the process and the physical construction of the reactor contribute to some unique temperature control problems. The nature of the chemical reaction results in a nonlinear process side demand on the control system. For example, some chemicals will react very rapidly, exhibiting a strong, initial exothermic reaction. The rate of reaction will decay exponentially with time, with the rate of decay being dependent on the characteristics of the particular process. Other processes will not exotherm until the reaction mass reaches a critical temperature or until a catalyst has been added. At times, the process side demand will be minimal, exhibiting the characteristics of a simple heat up to temperature and hold process much as would be found in heating a kettle of water. At other times, during the exotherm for example, the process side demand will be nonlinear because of reaction exotherm and will decay to the simpler heat up to temperature and hold type system. These widely varying changes in process side demand make selection and tuning of batch temperature controllers particularly critical.

The cyclical operation of the batch reaction cycle, as illustrated in Figure 7–4, is in itself nonlinear. Process operators will often load reactants into the reactor at room temperature, close the reactor for processing and set the batch temperature controller to the reaction temperature. Procedures may dictate that the batch come up to the control point without overpeaking. When pneumatic controllers are used, this requires the use of external feedback with accessory hardware to prevent reset windup in the controller. Electronic controllers utilize built-in diode limiters to eliminate the reset windup problem.

Batch process temperature controllers must be tuned for good startup performance (i.e., minimum time to reach the control point and no overpeak at the control point). Good control at the process set point is also required.

Application Data

Batch reactors range in size from 5 gallon (20 l) pilot plant reactors to full scale plant reactors holding 4000 gallons (15 000 l) or more. The choice of which batch reactor control system is dependent on the construction material of the reactor, the reaction cycle, and the heat transfer medium.

The reactor's physical construction, with its large reaction mass and relatively small heat transfer surface, causes the reaction mass to respond slowly to a change in controller set point. A period of oscillation of 45 minutes in a 1500 gallon (5 800 l) reactor is not uncommon. This makes it particularly critical to keep unnecessary process inputs from affecting the batch.

Thermal elements located in the batch are generally inserted in a well to provide mechanical rigidity and corrosion resistance. Thermal wells can add 15 to 20 seconds of lag to the control system. Heat transfer fluid in the well tip, speed-act response in the transmitter or preact response in the controller have been used to minimize this lag.

The temperature differential across the wall of a glass lined reactor must be limited to prevent damage to the glass. Special control systems have been developed to provide good process control while staying within the manufacturer's recommendations of maximum temperature differential.

CENTRIFUGAL COMPRESSOR SURGE CONTROL

The centrifugal compressor has become one of the most common means of gas compression used today. Its main advantage is mechanical simplicity compared to reciprocating and some other forms of rotary compressors. This generally is reflected in lower initial cost and lower maintenance cost. Each type of compressor exhibits application and control problems which are unique. Centrifugal compressor surge is unique and is a problem in most applications.

Surge is an unstable condition within the compressor blades. A centrifugal compressor head-flow curve is typified by the 100 percent speed line in Figure 7–6. As flow through the compressor decreases from maximum, the compression ratio increases to a maximum and falls off again with further flow decrease. This maximum ratio can occur near 50 percent of design flow for single stage units or as high as 70 percent for multistage units. When the flow through the compressor falls below this maximum pressure ratio point, the flow through the blades becomes unstable. This point is called the *surge point* of the compressor. For each compressor speed there is a characteristic head-flow curve with its characteristic surge point.

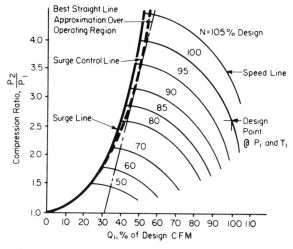

Figure 7–6. Typical compressor surge line data.

The line through the surge points at various compressor speeds is called the *surge line*, as shown in Figure 7–6.

CONTROL SYSTEM

Figure 7–6 is the typical form in which the centrifugal compressor manufacturer's surge data is usually presented. Alternately, this graph can be drawn from available data. The surge control line can then be plotted on this graph from 1 to 5 percent away from the actual surge line to provide protection from control system overshoot and recovery. The unstable flow condition in a centrifugal compressor is usually noisy and frequently violent enough to damage the compressor or associated piping. Therefore, in most compressor applications where the flow may drop below the surge line, it is important to provide a control system which will protect the compressor. In most cases where this is necessary, it is possible for the process to force the compressor into a surge condition very quickly. This means that the surge control system must be designed with maximum emphasis on response speed. This surge control line should then be replotted on graph coordinates which are more representative of the actual measured process variables, as shown in Figure 7–7.

On this graph, a straight line which is the best approximation of the surge control curve over the operating region can be plotted. This is most easily done by drawing a straight line through the points on the curve which represent the limits of the operating region and constructing a second straight line parallel to it which is tangent to the curve between these limit points.

For comparison, the best straight line approximation of the

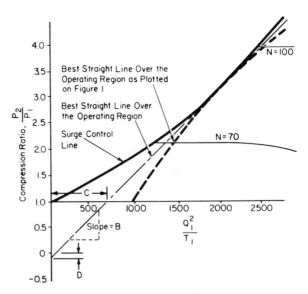

Figure 7–7. Surge control line replotted from Figure 7–6.

surge control line as plotted in Figure 7–6 is replotted in Figure 7–7. This shows that the straight line approximation of the surge control line as plotted on the coordinates in Figure 7–7 is closer to the actual surge control line than a straight line approximation made on the coordinates of Figure 7–6.

Since $Q_1 = \sqrt{K_1 h_1 T_1 P_1}$

Q_1 = suction flow

K_1 = proportionality constant

h_1 = suction line orifice differential pressure

T_1 = suction temperature

P_1 = suction pressure $\hspace{2cm}$ (1)

the x-axis in Figure 7–7, in terms of the actual measured processes variables, is

$$Q^2/T_1 = K_1{}^2 h_1/P_1 \hspace{2cm} (2)$$

Therefore, the equation of the straight line is

$$\frac{P_2}{P_1} = B(K_1{}^2 h_1/P_1) + D \hspace{2cm} (3)$$

P_2 = discharge pressure

B = slope of the straight line = $\dfrac{1 - D}{C}$

C = x-axis intercept

D = y-axis intercept

The process signal to the controller is the orifice plate differential pressure (h_1). All the other terms in equation (3) are used to compute the set point to this controller. Therefore, equation (3) should be rearranged into the following to represent the computed set point.

$$h_1 = \left(\frac{1}{BK_1{}^2}\right)(P_2 - DP_1) \hspace{2cm} (4)$$

The compressor control system which will hold the compressor flow at or above the straight line approximating the surge control line by using this equation is shown in Figure 7–8.

When the orifice plate differential pressure must be measured in the compressor discharge line, equation (4) must be changed to eliminate the h_1 term. Since the standard gas flow rate of the suction and discharge are equal, the following relationship can be used for this purpose.

$$Q_s = K_1\sqrt{h_1 P_1/T_1} = K_1\sqrt{h_2 P_2/T_2} \hspace{2cm} (5)$$

Q_s = flow

h_2 = discharge line orifice plate differential pressure

T_2 = temperature at discharge orifice plate

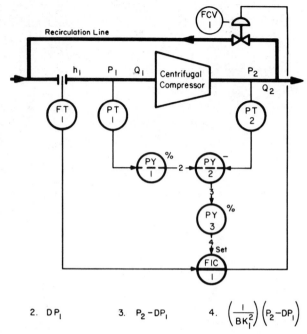

2. DP_1 3. $P_2 - DP_1$ 4. $\left(\dfrac{1}{BK_1^2}\right)\left(P_2 - DP_1\right)$

Figure 7–8. Compressor surge control system, orifice plate in suction line.

Equation (4) becomes:

$$h_2 = \left(\frac{1}{BK_1^2}\right)\left(\frac{T_2}{T_1}\right) P_1 \left(1 - D\frac{P_1}{P_2}\right) \qquad (6)$$

in terms of the discharge orifice plate differential pressure. The control system for this arrangement is shown in Figure 7–9.

Constant speed compressors are identical to variable speed compressors except that generally only a single speed line is shown on the manufacturer's data and the surge point is marked on this line. The Q_1 term in this case is a constant. Therefore, the equation for the controller set point is determined by rearranging equation (1) to the following form when the orifice plate is in the compressor suction line.

$$h_1 - (Q_1^2/K_1^2)\,(P_1/T_1) \qquad (7)$$

This control system is shown in Figure 7–10.

When the orifice plate is located in the discharge line of a constant speed compressor, equation (7) is changed to eliminate the h_1 term by again using equation (5). The equation for the surge controller set point then becomes:

$$h_2 = (Q_1^2 K_1^2)(P_1^2/P_2)(T_2/T_1^2) \qquad (8)$$

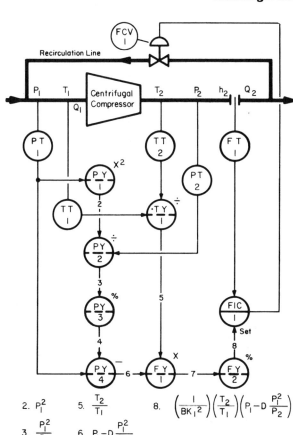

Figure 7–9. Compressor surge control system, orifice plate in discharge line.

2. P_1^2 5. $\dfrac{T_2}{T_1}$ 8. $\left(\dfrac{1}{BK_1^2}\right)\left(\dfrac{T_2}{T_1}\right)\left(P_1 - D\,\dfrac{P_1^2}{P_2}\right)$

3. $\dfrac{P_1^2}{P_2}$ 6. $P_1 - D\,\dfrac{P_1^2}{P_2}$

4. $D\,\dfrac{P_1^2}{P_2}$ 7. $\left(\dfrac{T_2}{T_1}\right)\left(P_1 - D\,\dfrac{P_1^2}{P_2}\right)$

The control system for this process configuration is shown in Figure 7–11.

Equations (4), (6), (7), and (8) can be simplified in cases where some of the variables are held constant. In addition to this, the D term in equations (4) and (6) can approach or equal zero, allowing the DP_1 or DP_1/P_2 term to be ignored.

Application Data

It should be strongly emphasized to anyone using a centrifugal compressor surge control system that the system is a protective device and, as such, is not to be adjusted as a plant operation variable. Therefore, all of the systems described are designed to permit manual adjustment of the surge controller set point when service of the more complex set point system is required. Adjustment of the system to a new surge control line or point can only be done by recomputation of the remote set point equation. Although this may

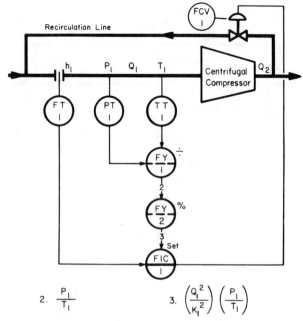

$$2. \quad \frac{P_1}{T_1} \qquad\qquad 3. \quad \left(\frac{Q_1^2}{K_1^2}\right)\left(\frac{P_1}{T_1}\right)$$

Figure 7–10. Fixed speed compressor control, orifice plate in suction line.

be inconvenient during the initial startup of the compressor, it is best for normal operation.

The combination of the compressor and the process is occasionally unstable. This condition is not compressor surge, although it may cause surge, and is due to misapplication of the compressor. It can only be corrected by changing the compressor or modifying the process.

INSTRUMENT LIST

	Tag No.	Description
Compressor surge control system with orifice plate in suction line	FT 1	Suction flow transmitter
Set-point computation components	PT 1	Suction pressure transmitter
	PT 2	Discharge pressure transmitter
	PY 1	Ratioing computer
	PY 2	Subtracting computer
	PY 3	Ratioing computer
Suction flow control system	FIC 1	Suction flow controller
	XY 1	Electro-pneumatic transducer (For use with valve on electronic systems)
	FCV 1	Suction flow control valve (recirculation flow)

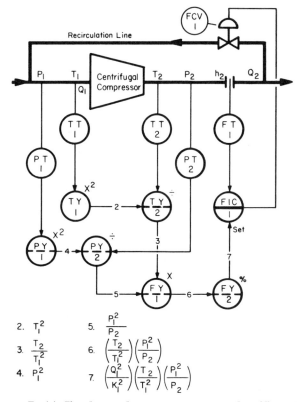

2. T_1^2

3. $\dfrac{T_2}{T_1^2}$

4. P_1^2

5. $\dfrac{P_1^2}{P_2}$

6. $\left(\dfrac{T_2}{T_1^2}\right)\left(\dfrac{P_1^2}{P_2}\right)$

7. $\left(\dfrac{Q_1^2}{K_1^2}\right)\left(\dfrac{T_2}{T_1^2}\right)\left(\dfrac{P_1^2}{P_2}\right)$

Figure 7–11. Fixed speed compressor control, orifice plate in discharge line.

	Tag No.	Description
Compressor surge control system with orifice plate in discharge line	FT 1	Discharge flow transmitter
Set-point computation components	PT 1	Suction pressure transmitter
	PT 2	Discharge pressure transmitter
	TT 1	Suction temperature transmitter
	TT 2	Discharge temperature transmitter
	TY 1	Dividing computer
	PY 1	Multiplying computer
	PY 2	Dividing computer
	PY 3	Ratioing computer
	PY 4	Subtracting computer
	FY 1	Multiplying computer
	FY 2	Ratioing computer

	Tag No.	Description
Discharge flow control system	FIC 1	Discharge flow controller
	XY 1	Electro-pneumatic transducer (For use with valve on electronic systems)
	FCV 1	Discharge flow control valve (recirculation flow)
Fixed speed compressor control with orifice plate in suction line	FT 1	Suction flow transmitter
Set-point computation components	PT 1	Suction pressure transmitter
	TT 1	Suction temperature transmitter
	FY 1	Dividing computer
	FY 2	Ratioing computer
Suction flow control system	FIC 1	Suction flow controller
	XY 1	Electro-pneumatic transducer (For use with valve on electronic systems)
	FCV 1	Suction flow control valve (recirculation flow)
Fixed speed compressor control with orifice plate on discharge line	FT 1	Discharge flow transmitter
Set-point computation components	PT 1	Suction pressure transmitter
	PT 2	Discharge pressure transmitter
	TT 1	Suction temperature transmitter
	TT 2	Discharge temperature transmitter
	PY 1	Multiplying computer
	PY 2	Dividing computer
	TY 1	Multiplying computer
	TY 2	Dividing computer
	FY 1	Multiplying computer
	FY 2	Ratioing computer
Discharge flow control system	FIC 1	Discharge flow controller
	XY 1	Electro-pneumatic transducer (For use with valve on electronic systems)
	FCV 1	Discharge flow control valve (recirculation flow)

BASIC RECIPROCATING COMPRESSOR CONTROL

Reciprocating compressors are the most common means of compressing gases. Their main feature is their high compression ratio per compression stage. This type of compressor can be driven by any type of prime mover such as an electric motor.

Figure 7–12 shows a typical reciprocating compressor cylinder. The rotary motion of the drive shaft is converted to the linear motion of the piston by the crank arm and the connecting rod. The suction and discharge check valves are actuated by the differential pressure across them. As the piston moves toward the head end of the cylinder, the gas in this end is compressed to the discharge pressure and forced out of the discharge check valve. At the same time, the pressure at the crank end of the cylinder has decreased and gas is drawn into the crank end of the cylinder through the suction check valve.

The suction and discharge check valves are truly check valves. They are, however, of a very special design which reduces the dead volume on the cylinder side of the valve to a minimum. This design allows them to operate for several years at a rate of several hundred times per minute. The compression ratio of the compression cylinder is the volume of the cylinder at maximum volume divided by the minimum volume. This ratio could be about 8 to 1. If the gas is drawn into the cylinder at 15 psig (100 kPag), the maximum pressure in the cylinder would be 120 psig (830 kPag). However, no gas could discharge at that pressure because the piston has reached its full travel. Therefore, the compressor is operated in the following manner. When the piston has reached 4/7ths (0.57) of its stroke, the pressure in the cylinder is 30 psig (210 kPag). The remaining 3/7ths (0.43) of the stroke discharges the gas at 30 psig (210 kPag). The volume of gas discharged is a function of the compression ratio, suction pressure, and the discharge pressure.

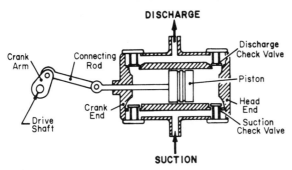

Figure 7–12. Typical reciprocating compressor cylinder.

CONTROL SYSTEM

The reciprocating compressor is basically an on/off device. Control is usually obtained by forcing the suction or discharge valves to stay open. In this condition, the pressure in the compressor cylinder is always at suction or discharge pressure. Figure 7–13 shows a typical arrangement for a suction valve actuator. When the actuating signal is applied to the diaphragm, the actuating fingers move against the valve plate, holding it away from the valve seat. The actuator shown unloads the compressor when air pressure is applied to the actuator. Air-to-load systems are also used. The configuration for a discharge unloader is basically the same. If it takes air to load the compressor, some other source of air pressure must be used to start the compressor. The pressure required to actuate the unloads may be from 30 psig (210 kPag) to full discharge pressure.

Figure 7–14 shows a two-step compressor unloading system used for discharge pressure control. The same equipment would be used for suction pressure control except that the controller action would be reversed. The discharge pressure is sensed by a differential gap type of controller. When the compressor is loaded, the discharge pressure rises. As it reaches the upper limit of the differential gap, the output pressure from the controller changes from one extreme level to the other to unload the compressor. Since the output pressure from the controller is not high enough or of great enough volume to actuate the unloader, a three-way valve is used to increase the pressure and volume.

When the compressor is unloaded, the discharge pressure will slowly decrease. At the low limit of the differential gap, the controller's output changes from one extreme to the other to load the compressor. The differential gap, instead of the usual on/off con-

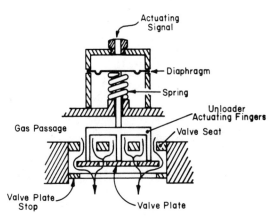

Figure 7–13. Typical suction valve with unloading actuator.

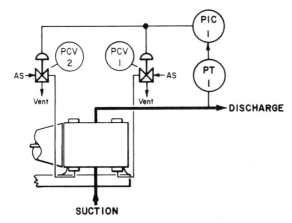

Figure 7—14. Two-step compressor unloading system with discharge pressure control.

troller, is used to prevent excessive wear on the unloader valves. An on/off controller would cause the valves to be actuated frequently. The volume of the discharge system, the compressor capacity, and the width of the differential gap determine the frequency of unloader valve actuation. Since this type of control permits the pressure to rise and fall noticeably, the compressor system must be designed to obtain the best balance between pressure variation and compressor wear.

A system which can control the compressor at half the variation in discharge pressure of the two-step system is shown in Figure 7–15. Two differential gap controllers are used in this system. The differential gap widths are the same for both controllers. The set points of the controllers are offset slightly so that one end of the compressor cylinder is loaded before the other as the pressure rises. The end which is loaded last is unloaded first. This usually results in less discharge pressure variation without increasing the wear on the compressor valves.

Application Data

The unloader valves and actuators are furnished by the compressor manufacturer. Whether the actuators are air-to-load or unload, the actuator operating pressures are determined by the compressor manufacturers. Because the crank shaft of the compressor must be designed to take certain mechanical forces, a compressor to take the desired type of unloading must be purchased. On multicylinder compressors, the manufacturer determines the combinations of head ends and crank ends that must be actuated together.

It is very important that the unloader actuators be operated as rapidly as possible. Slow actuation will cause the valves to chatter,

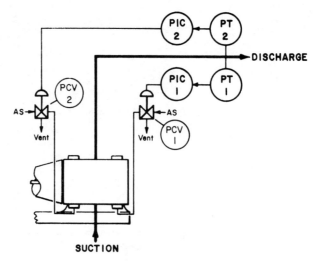

Figure 7–15. Three-step compressor unloading system with discharge pressure control.

causing excessive wear. Therefore, the three-way valves should be mounted on the compressor as close to the actuator as possible. The connecting tubing should be as large as possible—⅜ or ½ inch (9.5 or 12.7 mm) tubing is usually satisfactory.

INSTRUMENT LIST

	Tag No.	Description
Basic reciprocating compressor control system shown in Figure 7–14	PT 1	Pressure transmitter
	PIC 1	Pressure controller
	PCV 1	Three-way valve
	PCV 2	Three-way valve
Basic reciprocating compressor control system shown in Figure 7–15	PT 1	Pressure transmitter
	PT 2	Pressure transmitter
	PIC 1	Pressure controller
	PIC 2	Pressure controller
	PCV 1	Three-way valve
	PCV 2	Three-way valve

Index

Absorption oil process, 137–41
Absorption tower, 152
Activated sludge process, 87, 89–90
Actuation, pump, 108
Actuator, suction valve, 252, 253–54
Aeration process, 86–87, 89–90, 91
Air/fuel relationship, combustion control and, 6–8, 11–12
Alarm system, high temperature, 188
American Society for Testing Materials (ASTM), 142
Ammonia beer cooler control system, 37–41
Anion exchanger, 76, 77
Associated gas, 99
ASTM distillation, 142
Atmospheric cooling, retorting operation and, 55–59, 66, 68
Atmospheric distillation, 142–50
Atmospheric dye beck, 225–28
Attemperator, steam, 24–27

Balanced draft operation, 29
Batch
  digesters, 180
  flowmeter system, 182–83
  reactions, 238–43

Battery, central, 104, 109–14
Beam dye machine, 220–21
Beck, dye, 219–20, 225–28
Beer cooler control system, ammonia, 37–41
Biological oxygen demand, 85–86
Bleaching process, 201, 203
Blowback system, 183
Blowdown, 55–56, 58, 65
Boil control system for atmospheric dye becks, 225–28
Boilers, 1–31
  combustion, 1–16
  drum level control, 16–22
  furnace pressure control, 27–31
  steam temperature control, 23–27
  water demineralizing treatment for, 75–80
Boiling point, crude oil distillation, 142, 145
Booster station control, pipeline, 162–70
Brewery, 37–53
  ammonia beer cooler control system, 37–41
  carbonation control system for, 41–49
  clean-in-place system for, 50–53

Brine density control system, 33–34
Bubble point, 152

Calcium hypochlorite bleach process control system, 203, 205
Calender, 215–17
Calender rollers, 216
Canning. *See* Retorting operation
Carbonation control system for breweries, 41–49
Case pressure, 165
Casing, well, 101–3
Cation exchanger, 76, 77
Central battery, 104, 109–14
Centrifugal compressor surge control, 243–50
Centrifugal sludge dewatering, 95
Charging system, 183
Chemical addition, water treatment by, 71–73
Chemical digestion process, 178
Chemical oxygen demand, 85–86
Chemical reactors, 237–43
Chemicals, emulsion-breaking, 113
Chip cooker, 190, 194–95
Chip screen, 171–73
Chlorination process, 74–75
Choke, 104–5
Clarification, sludge, 92–93
Clarification, water, 71–73
Clean-in-place system, 49–53
CO_2. *See also* Carbonation
flow rate, 45
removal process, 133–34, 135
volumes measurement, 44–45
Coagulation process, 73

Combination feedforward-feedback system, 19–20
Combustibles analysis, 14–15, 16
Combustion
in boilers, 1–16
efficiency, 2–3, 4, 6–7, 11–14
furnace pressure operation, 28–29
heater, 145–47
Compressor control, 243–54
centrifugal, 243–50
reciprocating, 251–54
Computer, feedwater flow, 20
Condensate separator, 124, 125
Condenser accumulator water temperature system, 188
Conditioning process, pulp, 198–207
Conical screw, 190, 192, 194, 197
Contactor control system, 132
Continuous reactions, 239
Controller
logic, 77, 78
oxidation-reduction potential, 203–6
ramp set point, 166–68
Convection type superheater, 23
Conveyor screw, 171–76, 190, 192–97
Cooker, chip, 190, 194–95
Cooking
liquor measuring systems, 182–85
retorting operation, 54–70
—steam, 54–64
—water, 54, 65–70
in sulfate chemical digestion, 178–80
Cooler control system, beer, 37–41
Cooling period in retorting operation, 55–59, 66, 68

Crude oil
    assay, 142
    distillation, 142–51
    from LACT unit, 117
    lines, 157
    in mechanical lift well, 108
Custody transfer unit, lease
    automatic, 114–17, 125
Cut point, 142
Cyclone, 195
Cyclone cleaners, 200, 203, 204

Dead time of CO_2 volumes
    measurement, 44–45
Degasser, 76, 77–79, 80
Demineralization water
    treatment, 75–80
Density control system
    brine, 33–34
    lye, 34–35
Deslater level, 143
Dewatering, sludge, 93–96
Diethanolamine, gas treatment
    with, 133
Diethylene glycol, gas
    treatment with, 129
Digester, sulfate chemical,
    178–82
Digester blow heat recovery,
    186–89
Digestion, sludge, 92
Dilution, combustion
    efficiency and, 3, 13
Dilution water in
    thermomechanical
    pulping, 195, 197
Direct contact attemperator, 24
Direct steaming, 180
Discharge check valves, 251,
    252
Dissolved gas, 99
Dissolved oxygen sensor, 89–
    90
Distillation, crude oil, 142–51

Draft operation of furnace, 28–
    29
Drier stage, paper machine,
    212–14
Drilling, oil and gas, 100–101
Drug addition in textile
    dyeing, 224
Drum filters, 94–95, 200–201,
    203, 204
Drum level control, boiler, 16–
    22
Dry end, paper machine, 207–
    8
Dry gas, 99
Dry hole, 101
Dye beck, 219–20, 225–28
Dyeing of textiles, 219–25
Dye jig, 222
Dynamometer card, 109

Effluent water treatment, 80–
    91
Emulsion-breaking chemicals,
    113
Endless screw conveyor, 174–
    76, 192
Endothermic reactions, 238
End point, 142
Energy content of fuels, 1
Exothermic reactions, 238

Faults in rock formations, 100,
    101
Feedback approach to
    combustion control, 8–9
Feedforward, 149, 150, 157
Feedwater flow, 20, 21
Film processing, 234–37
Filtration
    drum filters, 94–95, 200–201,
        203, 204
    sand filters, 73–74
    sludge dewatering, 93, 94–95
    trickling, process, 87–88
Firing control, rate of, 4, 5–6,
    11

Flash tank control system, 133
Flash vaporization, 137
Flash zone, 142
Flocculation process, 73
Flocculent addition, 83
Flotation, 81–82, 83–84, 93
Fluidized bed incineration, 97–98
Food industry, 33–70
  ammonia beer cooler control system, 37–41
  brine density control system, 33–34
  carbonation control system, 41–49
  clean-in-place system, 49–53
  high temperature short-time pasteurization, 35–37
  lye density control system, 34–35
  retorting operation, 53–70
Forced draft operation, 29
Fractionation, light ends, 153–55
Fuels. See Combustion; Petroleum
Furnace pressure control, 27–31. See also Boilers

Gas. See also Petroleum
  heater, 123–25
  production, 99–103, 122–28
  scrubbing tower, 231–33
  -to-gathering system, 125–26
  treatment, 128–36
  types of, 99
Gathering system, 122–23, 125–26
Generator, ramp function, 165–66, 167, 169
Gravity filtration sludge dewatering, 93

H₂S removal process, 133–34, 135

Head box, 208–9, 210
Heater(s)
  combustion, 145–47
  controls, 121
  gas, 123–25
  liquor, 178
  oil, 118–19
  pass flow, 145
Heat recovery, digester blow, 186–89
Heat recovery in gas scrubbing tower, 231–33
Heat transfer, 240–41
Heavy key component, 152
Helium removal from gas, 135
High temperature alarm system, 188
High temperature short-time pasteurization, 35–37
Horizontal retort, 54, 62–63, 65, 69
Hot vapor bypass, 153, 155
Hot water storage tank level system, 188–89
Hot water storage temperature system, 188
Hydrocarbons, 99–100, 152, 153. See also Petroleum

Ignition temperature, 2
Incineration, 96–98
Indirect steaming, 178–80
Induced draft operation, 28
Influent metering, 84
Influent water treatment, 71–75
Initial boiling point, 142
Injection type head box, 208–9
Isomer, 152

Jacketed batch reactor, 239–40
Jet dye machine, 223–24

Landfill, 96
Lean absorption oil, 137, 138

Lean oil, 152
Lean oil split, 138–40
Lease automatic custody transfer (LACT) unit, 114–17
Light ends recovery, 151–57
Light key component, 152
Liquor heater, 178
Liquor measuring systems, cooking, 182–85
Loading surge control, 88–89
Logic controller, 77, 78
Lye density control system, 34–35

Main valves, 117
Manifold purging, 112
Mechanical lift wells, 106–9
Metering, influent, 84
Milk pasteurization, 35–37
Monitoring, pump, 108–9
Monoethanolamine, gas treatment with, 133–34, 135
Multiple completion well, 101, 102
Multiple hearth incineration, 96–97

Natural draft operation, 28
Natural gas lines, 157
Natural lift wells, 103–6
Nitrogen removal from gas, 135
Nonassociated gas production, 122–28

Oil. See also Petroleum
    absorption process, 137–41
    crude, 108, 117, 142–51, 157
    heater, 118–19
    lean, 138–40, 152
    production, 99–103
    rich, 137, 138–41
    sponge, 137
Organic material trap, 76

Orifice plate differential pressure, 245–47
Overflash, 142
Overhead vapor composition, 155
Overhead vapor temperature, 148–49, 150
Oxidation-reduction potential controller, 203–6
Oxygen
    demand, 85–86
    sensor, dissolved, 89–90
    trim control system, 13–16
Ozonation process, 74–75

Package dye machine, 221–22
Padder, 222
Paddle dye machine, 220
Paper. See Pulp and paper
Pasteurization, 35–37
Peeler lye density control system, 34–35
Petroleum, 99–170
    absorption oil process, 137–41
    central battery, 104, 109–14
    crude oil distillation, 142–51
    gas treatment, 128–36
    lease automatic custody transfer unit, 114–17
    light ends recovery, 151–57
    mechanical lift wells, 106–9
    natural lift wells, 103–6
    oil and gas production, 99–103, 122–28
    pipeline booster station control, 162–70
    pipeline operations, basic, 157–62
    separators, 110–13, 114, 117–22, 124, 125
pH adjustment, 83–84
pH control system, 88–89
Piece dye beck, 220
Pipeline, See under Petroleum

Piping, well, 101–3
Polished rod, 108
Pollution, water. *See* Water
    treatment
Positive displacement pumps,
    169
Prechlorination, 75
Preheat period in retorting
    operation, 65
Presaturation, 137
Pressure
    case, 165
    cool retort control system,
        steam cook, 60–64
    dye vessel, control system,
        224
    furnace, control, 27–31
    orifice plate differential,
        245–47
    pipeline, 158–60
    protection systems, 126–27
    retort, 55–56, 58, 61–63, 65,
        68–69
Primary waste treatment, 80–
    84
Production separator, 110–13,
    114
Product lines, 157–58
Protection from overpressure
    system, 126–27
Pulp and paper, 171–217
    conditioning process, 198–
        207
    cooking liquor measuring
        systems, 182–85
    digester blow heat recovery,
        186–89
    paper machine, 207–17
    sulfate chemical digester,
        178–82
    thermomechanical pulping,
        190–98
    wood chip handling, 171–77
Pump, positive displacement,
    169. *See also* Wells

Pumparound, 142, 147, 148,
    150
Pumpback, 150

Radiant type superheater, 23
Ramp function generator, 165–
    66, 167, 169
Ramp set point controller, 166–
    68
Reactor control, 237–43
Reciprocating compressor
    control, 251–54
Recorder, three-pen, 47
Refining in thermomechanical
    pulping, 190–98
Reflux, 152
Regenerator, 129–31, 133
Relief control system, 181
Replenishment loops in film
    processing, 234–35
Residue gas, 137
Retorting operation, 53–70
    steam cooking, 54–64
    water cooking, 54, 65–70
Rich oil, 137, 138
Rich oil demethanizing, 138–
    41
Rollers, paper machine, 207,
    210–17

Safety limits, booster station,
    163–64
Sand filters, 73–74
Saturated hydrocarbons, 152
Saturates gas plant, 153, 154
Screens
    chip, 171–73
    primary treatment, 82–83
    pulp, 198, 201–2
Screw
    conical, 190, 192, 194, 197
    conveyor, 171–76, 190, 192–
        97
    press, 198, 200

Scrubbing tower, gas, 231–33
Secondary waste treatment, 84–91
Sedimentation, 18, 82, 73, 83–84
Separators, 110–13, 114, 117–22, 124, 125
Sheet former, 210
Side-cut product draw, 147–48, 149
Single element drum level control system, 17–19
Single element attemperator system, 25, 26
Single element furnace draft control, 29–31
Skein dye machine, 223
Slasher, high speed warp, 228–30
Sludge
 concentration, 92–96
 disposal, 96–98
 process, activated, 87, 89–90
Splitter, 153
Sponge oil, 137
Spray attemperator, 24–25
Stabilizer, 153
Startup, booster station, 165
Station operation, pipeline, 161–70
Steam
 cooking, retorting operation for, 54–64
 flow measurement, 19
 header pressure, 5–6, 11, 16
 temperature control, 23–27
Steaming in sulfate chemical digestion, 178–81
Still, rich oil, 137, 140
Stress, pump, 109
Stripper, 153
Stripper bottoms temperature, 155
Sucker rod, 107–8
Suction check valve, 251, 252

Suction-discharge override control system, 163
Suction valve actuator, 252, 253–54
Surge control, loading, 88–89
Sulfate chemical digester, 178–82
Superheaters, 23, 24–25. See also Boilers
Supply side load changes, 8
Surface attemperator, 24

Temperature
 in digester blow heat recovery, 188
 dye liquor, 224, 225
 ignition, 2
 overhead vapor, 148–49, 150
 slasher cylinder, 229
 steam, control, 23–27
 stripper bottoms, 155
Test separator, 110–13, 114
Textiles, 219–30
 boil control systems for atmospheric dye becks, 225–28
 dyeing of yarn and fabric, 219–25
 high speed warp splasher, 228–30
Thermomechanical pulping, 190–98
Three element drum level control system, 21–22
Three-pen recorder, 47
Three pinch roller press, 207, 210–12
Time delay relay, 44
Time-temperature profile for dyeing, 224, 225
Total organic carbon, 85–86
Tower bottoms composition, 157
Traps, oil and gas, 100, 101
Traps, organic material, 76

Traveling valve, 107
Trickling filtration process, 87–88, 90, 91
Triethylene glycol, gas drying with, 129–31
True boiling point distillation, 142
Two element drum level system, 19–20

Unsaturated hydrocarbons, 153
Unsaturates gas plant, 153

Vacuum distillation, 149–50, 151
Vacuum filtration sludge dewatering, 94–95
Valve(s)
  check, 251, 252
  drum level control, 18–19
  main, 117
  operation in mechanical lift wells, 107
  steam temperature control, 27
  suction, 252, 253–54
  traveling, 107
  wing, 103
Vapor bypass, hot, 153, 155
Vaporization, flash, 137
Vapor temperature, overhead, 148–49, 150
Venting period in retorting operation, 55, 57, 60–61

Vertical retort, 54, 61–62, 65, 69

Warp slasher, high speed, 228–30
Wastewater purification process, 80–91
Water cooking, retorting operation for, 54, 65–70
Water removal from gas, 129–31
Water treatment, 71–98
  demineralizing, 75–80
  effluent, 80–91
  influent, 71–75
  sludge concentration, 92–96
  sludge disposal, 96–98
Wells
  casing and piping, 101–3
  control, 126
  drilling oil and gas, 100–101
  location, steam temperature control and, 25–27
  mechanical lift, 106–9
  natural lift, 103–6
  shut-in, 126
Wet end, paper machine, 207
Wet gas, 99
Wet-well level control, 88–89, 90
Wing valve, 103
Wood chip handling, 171–77. See also Pulp and paper